Design, Fabrication and Performance of Wind Turbines 2020

Design, Fabrication and Performance of Wind Turbines 2020

Editor

Kyung Chun Kim

MDPI • Basel • Beijing • Wuhan • Barcelona • Belgrade • Manchester • Tokyo • Cluj • Tianjin

Editor
Kyung Chun Kim
School of Mechanical Engineering,
Pusan National University
Korea

Editorial Office
MDPI
St. Alban-Anlage 66
4052 Basel, Switzerland

This is a reprint of articles from the Special Issue published online in the open access journal *Energies* (ISSN 1996-1073) (available at: https://www.mdpi.com/journal/energies/special_issues/DFP_WT2020).

For citation purposes, cite each article independently as indicated on the article page online and as indicated below:

LastName, A.A.; LastName, B.B.; LastName, C.C. Article Title. *Journal Name* **Year**, *Volume Number*, Page Range.

ISBN 978-3-0365-0314-1 (Hbk)
ISBN 978-3-0365-0315-8 (PDF)

© 2021 by the authors. Articles in this book are Open Access and distributed under the Creative Commons Attribution (CC BY) license, which allows users to download, copy and build upon published articles, as long as the author and publisher are properly credited, which ensures maximum dissemination and a wider impact of our publications.

The book as a whole is distributed by MDPI under the terms and conditions of the Creative Commons license CC BY-NC-ND.

Contents

About the Editor vii

Preface to "Design, Fabrication and Performance of Wind Turbines 2020" ix

Yuan Song and Insu Paek
Prediction and Validation of the Annual Energy Production of a Wind Turbine Using WindSim
and a Dynamic Wind Turbine Model
Reprinted from: *Energies* **2020**, *13*, 6604, doi:10.3390/en13246604 1

Kyoungboo Yang
Geometry Design Optimization of a Wind Turbine Blade Considering Effects on Aerodynamic
Performance by Linearization
Reprinted from: *Energies* **2020**, *13*, 2320, doi:10.3390/en13092320 17

Enyu Cai, Yunqiang Yan, Lei Dong and Xiaozhong Liao
A Control Scheme with the Variable-Speed Pitch System for Wind Turbines during a
Zero-Voltage Ride Through
Reprinted from: *Energies* **2020**, *13*, 3344, doi:10.3390/en13133344 35

Shyh-Kuang Ueng
A Hybrid RCS Reduction Method for Wind Turbines
Reprinted from: *Energies* **2020**, *13*, 5078, doi:10.3390/en13195078 57

Mihai Chirca, Marius Dranca, Claudiu Alexandru Oprea, Petre-Dorel Teodosescu, Alexandru Madalin Pacuraru, Calin Neamtu and Stefan Breban
Electronically Controlled Actuators for a Micro Wind Turbine Furling Mechanism [†]
Reprinted from: *Energies* **2020**, *13*, 4207, doi:10.3390/en13164207 71

Wei Li, Leilei Ji, Weidong Shi, Ling Zhou, Hao Chang and Ramesh K. Agarwal
Expansion of High Efficiency Region of Wind Energy Centrifugal Pump Based on Factorial
Experiment Design and Computational Fluid Dynamics
Reprinted from: *Energies* **2020**, *13*, 483, doi:10.3390/en13020483 85

About the Editor

Kyung Chun Kim is a distinguished professor at the School of Mechanical Engineering of Pusan National University in Korea. He obtained his Ph.D. from the Korea Advanced Institute of Science and Technology (KAIST), Korea, in 1987. He was selected as a member of the National Academy of Engineering of Korea in 2004. His research interests include flow measurements based on PIV/LIF, turbulence, heat transfer, organic rankine cycle, wind turbine and fuel cell, and wind engineering.

Preface to "Design, Fabrication and Performance of Wind Turbines 2020"

The consumption of fossil fuels has increased, resulting in high CO2 emissions and serious climate change. Research on renewable energy is currently underway in order to solve these environmental problems, and in anticipation of the depletion of fossil fuels. Wind energy is an environmentally friendly renewable energy source that does not cause environmental pollution, and its use is rapidly spreading around the world. From small-scale vertical axis wind turbines for urban usage to large-scale horizontal axis wind turbines for offshore wind farms, design, fabrication, and optimization technologies are highly required to manage wind energy effectively. Moreover, some new methods, such as for wind farm design, fluid–structure interaction, aero-acoustics, fabrication methods and performance tests by experimental and computational fluid dynamics should be implemented in modern wind turbine communities. Our basic objectives include improving the reliability, promoting the high efficiency of wind turbines, dynamic performance, reducing wind turbine generated noise and improving power generation efficiencies through high-fidelity approaches. Managing such a wide range of wind turbine scales and usages, design, fabrication, and performance test protocols for various wind turbines is a challenging issue. This Special Issue aims at encouraging researchers to address solutions to overcome the issue.

Kyung Chun Kim
Editor

Article

Prediction and Validation of the Annual Energy Production of a Wind Turbine Using WindSim and a Dynamic Wind Turbine Model

Yuan Song [1] and Insu Paek [2,3,*]

[1] Department of Advanced Mechanical Engineering, Kangwon National University, Chuncheon 24341, Gangwon, Korea; songwon@kangwon.ac.kr
[2] Department of Integrated Energy & Infra System, Kangwon National University, Chuncheon 24341, Gangwon, Korea
[3] Department of Mechatronics Engineering, Kangwon National University, Chuncheon 24341, Gangwon, Korea
* Correspondence: paek@kangwon.ac.kr

Received: 4 August 2020; Accepted: 11 December 2020; Published: 14 December 2020

Abstract: In this study, dynamic simulations of a wind turbine were performed to predict its dynamic performance, and the results were experimentally validated. The dynamic simulation received time-domain wind speed and direction data and predicted the power output by applying control algorithms. The target wind turbine for the simulation was a 2 MW wind turbine installed in an onshore wind farm. The wind speed and direction data for the simulation were obtained from WindSim, which is a commercial computational fluid dynamics (CFD) code for wind farm design, and measured wind speed and direction data with a mast were used for WindSim. For the simulation, the wind turbine controller was tuned to match the power curve of the target wind turbine. The dynamic simulation was performed for a period of one year, and the results were compared with the results from WindSim and the measurement. It was found from the comparison that the annual energy production (AEP) of a wind turbine can be accurately predicted using a dynamic wind turbine model with a controller that takes into account both power regulations and yaw actions with wind speed and direction data obtained from WindSim.

Keywords: onshore wind farm; flow analysis; annual energy production; dynamic wind turbine model; yaw motion; peak shaver

1. Introduction

South Korea has recently been actively promoting the development of power plants using renewable energy. At the end of 2017, the Ministry of Trade, Industry, and Energy of South Korea issued the "Renewable Energy 3020 Plan", which aims to achieve 20% of the total power generation in South Korea from renewable energy by 2030. The 20% target is approximately 63.8 GW in capacity, about 28% of which, approximately 17.7 GW, corresponds to wind power. Prior to 2019, South Korea's cumulative total of wind power was approximately 1.49 GW, which is about 8.4% of the anticipated renewable energy production in the 3020 plan [1,2]. To achieve the target of the "Renewable Energy 3020 Plan", it is necessary to substantially develop more wind farms and investigate more wind farm sites based on accurate prediction of the annual energy production (AEP), which must be performed in advance.

To predict the AEP of a wind farm at a specific site, the wind data representing that site are needed. The wind data can be either time-based or frequency-based (Weibull distribution). An appropriate program is needed to predict the wind speeds at different wind turbine locations in the wind farm, including the wake effect, and also to predict the AEP from the wind turbines. Programs that are used to predict the AEP of a wind farm are already commercialized and are used worldwide. The programs are classified as linear and nonlinear (CFD) programs based on the equation of motion. The linear programs include WAsP, WindPRO, and WindFarmer, and the nonlinear programs include WindSim, WAsP-CFD, Meteodyn, and Wind Station.

Although the linear programs are known to predict the AEPs of a wind farm well for flat terrains, they have limitations when used for complex terrains. With the increase in computing speed and storage capacity, analysis with CFD is being used more frequently than before to obtain more accurate wind resources.

Some studies have predicted wind resources of various regions to find suitable wind farm sites using commercial CFD codes [3–6]. Dhunny et al. used WindSim to predict wind resources for the island of Mauritius [3]. Similar studies on finding suitable offshore wind farm sites using the CFD simulation WindSim can be found in Yue et al. [4] and Song et al. [5]. Park et al. used WindSim to predict wind resources of wind farms in complex terrains and to find suitable offshore wind farm sites [6]. However, no experimental validations of the simulation results were made in these studies.

There are also studies predicting the AEPs of operating wind farms using commercial software (such as WindPRO, WindSim, WindFarm, and Metodyn) with either measured or reanalyzed wind data and validating the results with actual energy productions. Tabas et al. used WindSim to predict the power of wind by using different turbulence and wake models and compared the results with the measured power. In their study, the root mean square errors were between 0.09% and 0.25% [7]. Kim et al. used WindSim to predict the AEPs of two wind farms on complex terrain in South Korea, and the errors compared with the actual AEPs for three years were between 1.7% and 10.9% [8]. Song et al. used WindSim and WindPRO to predict the AEPs of an offshore wind farm in Denmark, and the errors compared with the actual AEPs for three years were between 0.56% and 4.64% and between 0.09% and 5.71%, respectively [9]. Kim et al. used WindPRO to predict the AEP of a small wind farm with two turbines by considering nearby obstacles. The errors between the prediction and the measured AEP for three years were between 0.9% and 2.4% [10]. Results from these studies indicate that the errors in the AEP prediction of wind farms were mostly within 10% [7–10].

The commercial software used to predict the electrical power produced by a wind turbine is generally based on the Weibull distribution of the wind data and the steady power curve of the wind turbine, so the wind turbine orientation is always matched with the wind direction, and the power output is based on the steady power output. Therefore, the wind turbine yaw dynamics and the control algorithm to face the wind during wind direction changes are not considered, and the transient power variation in actual operating conditions is not simulated. The steady power curve is also based on the wind turbine control, which is applied when the power curve of the wind turbine is made, and therefore, any changes in the operating points of the wind turbine from the steady power curve cannot be simulated.

However, there is increasing interest in research on wind farm control, which actively adjusts the power demand distribution or yaw demand distribution to individual wind turbines in a wind farm to maximize the total power output from the wind farm [11–14]. For this kind of simulation, the wind turbine must be modeled for dynamic simulation, and it must have suitable power control algorithms so that the power output from the wind turbine varies with the commands from wind farm controllers.

A recent study proposed using a dynamic wind turbine simulation model to predict the AEP of a wind turbine [15]. It used a Simulink version of the NREL 5 MW paper wind turbine including a controller to predict the AEP of a wind turbine at four different sites and compared the results with the results from a commercial code, WindPRO. The predictions from the dynamic simulation code were found to be a few percent smaller than those from the commercial code due to yaw motion. However, the results were not experimentally validated in that study.

Therefore, in this study, the previous study by Kim et al. was revisited to experimentally validate the AEP predicted by the model [15]. For this, the dynamic Simulink model was used to model an actual 2 MW wind turbine that is implemented in a wind farm in Korea. The wind turbine control was also tuned to produce the power curve of the wind turbine, and the revised wind turbine model was used to predict the annual energy production of the wind turbine. The wind data for the wind turbine were obtained from a commercial wind farm simulation code, WindSim, with the wind data measured from a meteorological mast near the wind turbine. Finally, the prediction results from the dynamic simulation code were compared with the results from the commercial code, WindSim, and the measurement.

2. Methodology

2.1. Target Wind Turbine

In this study, a wind turbine in a commercial wind farm D in South Korea was used as a target wind turbine. As shown in Figure 1, wind farm D consists of 15 wind turbines. The met mast is at the north of the target wind farm. The target wind turbine is about 2.5 RD (Rotor Diameter) away from the met mast, which corresponds to 220 m in distance. The met mast and the target wind turbine are located in a relatively flat terrain, and they are about 1.7 km away from the sea.

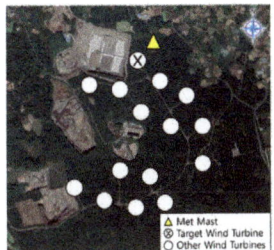

Figure 1. The layout of wind farm D (Source: Google Maps).

Table 1 shows the general specification of the target wind turbine. As shown in the table, the hub height of the wind turbine is 80 m. The cut-in, rated, cut-out wind speeds are 3.5 m/s, 12 m/s, and 25 m/s, respectively. Figure 2 shows the power curve showing the power output with respect to the wind speed for the standard air density, which is 1.225 kg/m^3.

Table 1. General specification of the target wind turbine.

Properties	Value
Rated power	2 MW
Cut-in, rated, cut-out Wind speed	3.5 m/s, 12 m/s, 25 m/s
Hub height	80 m
Power control	Pitch Control

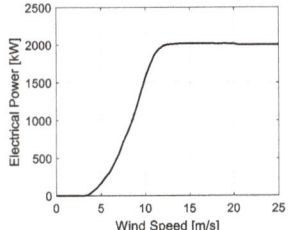

Figure 2. The power curve of the target wind turbine.

2.2. Wind Data

In this study, the 10-min averaged data measured from 1 January 2017 to 31 December 2017 at the meteorological mast were used. The meteorological mast sensors comprise anemometers, wind direction vanes, a temperature sensor, a barometric pressure sensor, and a relative humidity sensor. The anemometers were installed at heights of 80 m, 78 m, and 40 m. The wind direction vanes were installed at heights of 78 m and 40 m. The temperature sensor, barometric pressure sensor, and relative humidity sensor were installed at a height of 75 m. The data measured with the 80 m anemometer and 78 m wind direction vane were used to predict the electrical power production of the target wind turbine in the wind farm. Figure 3a shows the Weibull distribution, and Figure 3b is the wind energy rose obtained using the met mast data. As shown in Figure 3, the average wind speed at a height of 80 m is 5.75 m/s, and the prevailing wind direction is 315 degrees in 16 sector-wise angles.

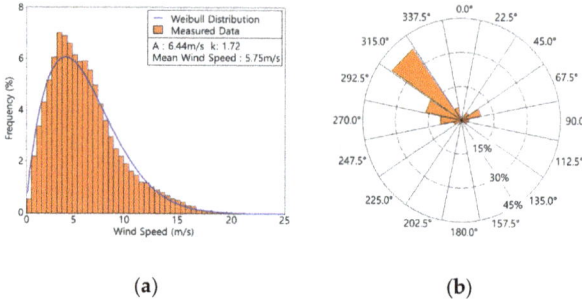

(a)   (b)

Figure 3. Measured wind data for the target wind turbine: (**a**) Weibull distribution; (**b**) total wind energy rose.

According to the equation for data normalization provided by the International Electrotechnical Commission (IEC) standards 61400-12-1 2nd edition [16], the air density at the target wind turbine can be calculated using the measured 10-min average data for temperature and air pressure of the mast at a height of 75 m. The 10-min average air density was calculated using Equation (1):

$$\rho_{10min} = \frac{B_{10min}}{R_0 \times T_{10min}}. \tag{1}$$

In the equation, ρ_{10min} is the calculated 10-min average air density, B_{10min} is the measured 10-min average air pressure, T_{10min} is the measured 10-min average air temperature, and R_0 is the gas constant of dry air, which is 287.05 J/(kg·K). The air density calculated at a height of 75 m at the met mast position is 1.076 kg/m³. The air density for the target wind turbine was assumed to be the same as the air density at the met mast.

2.3. Terrain Modeling

A terrain model that includes the information on both the topology and roughness was used for the CFD simulation analysis of the wind farm. The height contour map provided by the Korea National Spatial Information Portal and the land cover data provided by the Korea Ministry of Environment [17,18] were used to construct the terrain model.

The terrain model of the target wind farm was constructed by the geographic information system (GIS) software Global Mapper based on the height contour and the land cover data and implemented with the CFD simulation program, WindSim. Figure 4a,b shows the elevation and the roughness length information of the terrain model.

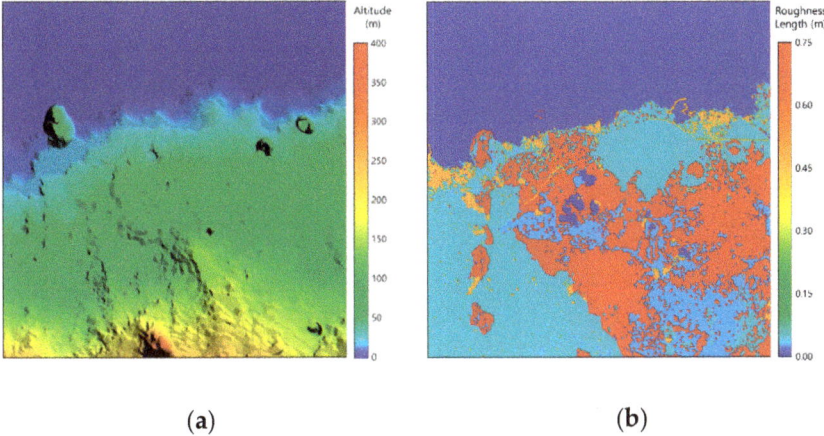

Figure 4. Terrain model of the target wind farm L (**a**) elevation; (**b**) roughness length.

The size and grid spacing of the terrain model has an impact on the prediction results of the power output [19]. Therefore, in this study, the size of the terrain model was determined to be 10 km by 10 km, and the horizontal grid spacing was 25 m by 25 m. The number of vertical grids was 40, and the number of grids below the wind turbine hub height was six, to accurately simulate the wind characteristics in the wind farm.

The wind field analysis was carried out using WindSim based on the Reynolds-Averaged Navier–Stokes (RANS) equation. Because the equation is nonlinear, the solution is solved with the use of multiple iterations until convergence is obtained [20]. This study uses the GCV (Convergence control with new solver) solver in the CFD simulation WindSim to solve the RANS equation. The GCV method uses a pressure-based separation solver strategy to control the mass flux on the control-volume faces, providing faster convergence [21]. It was assumed that the atmospheric boundary layer had a height of 500 m, that the atmospheric flow was stable, and that the wind speed above 500 m was constant with height [22]. The boundary condition at the top was "fixed pressure". The turbulence model used was the standard k-ε eddy viscosity turbulence model without considering temperature changes, where k is the turbulent flow energy and ε is the rate of turbulent dissipation.

Based on the results of the wind field analysis and the N.O. Jensen wake model [23], the 10-min averaged wind speed, wind direction, and electrical power of the target wind turbine were obtained from WindSim.

2.4. The Wind Turbine Matlab/Simulink Dynamic Model with a Wind Turbine Controller

In this study, the dynamic simulation model of a wind turbine was used for the prediction of the electrical power production by the target wind turbine. The model was developed in the previous study and modified for this study to be used to predict the power of an actual wind turbine [15].

Figure 5 shows the schematic of the dynamic model used for the target wind turbine. The dynamic model of the wind turbine is composed of four parts: the wind turbulence model, the yaw system model, the wind turbine controller, and the wind turbine model. The model requires 10-min average wind speed and direction data together with their standard deviations as inputs and calculates the electrical power output of the wind turbine with the power and yaw control algorithms. Detailed information on the model is available in Ref. [15].

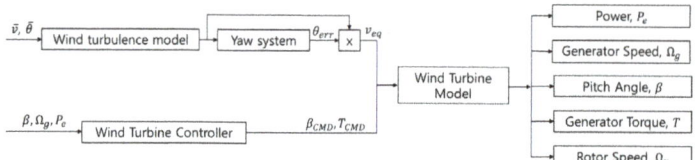

Figure 5. The wind turbine dynamic model.

2.4.1. The Wind Turbine Matlab/Simulink Dynamic Model

In this study, the wind turbine dynamic model was simulated by using Matlab/Simulink, and the block diagram of the dynamic wind turbine model is shown in Figure 6. As shown in the figure, the wind turbine dynamic model is composed of the aerodynamics model, the drive train model, the generator model, the pitch actuator model, and the control system model.

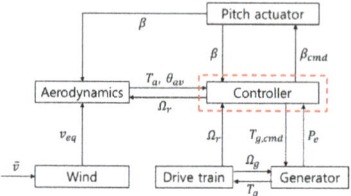

Figure 6. The block diagram of the wind turbine model.

The aerodynamic power can be simply modeled using Equation (2).

$$P_{aero} = \frac{1}{2}\rho\pi R^2 C_P(\lambda, \beta) v^3, \quad (2)$$

where P_{aero} is the aerodynamics power of the wind turbine, ρ is the air density, R is the wind turbine radius, v is the wind speed, and C_P is the power coefficient of the wind turbine which depends on the tip speed ratio (λ) and the blade pitch angle (β). The power coefficients for various tip speed ratios and blade pitch angles are calculated from a BEMT (Blade Element Momentum Theory) software such as Bladed and used as a look-up table in the aerodynamics model.

The drive train model is composed of a low-speed shaft, a gearbox, and a high-speed shaft. The aerodynamics torque and generator torque are the inputs of the drive train model. The low-speed shaft and high-speed shaft can be modeled by the inertia, the stiffness, and the damping. The drive train model was simulated using Equations (3) and (4). The relationship between the low-speed shaft and the high-speed shaft is shown in Equation (5) [24].

$$J_r \frac{d\Omega_r}{dt} = T_a - k_s\left(\theta_r - \frac{1}{N}\theta_g\right) - c_s\left(\Omega_r - \frac{1}{N}\Omega_g\right) - B_r\Omega_g, \quad (3)$$

$$J_g \frac{d\Omega_g}{dt} = \frac{k_s}{N}\left(\theta_r - \frac{1}{N}\theta_g\right) + \frac{c_s}{N}\left(\Omega_r - \frac{1}{N}\Omega_g\right) - B_g\Omega_g - T_g, \quad (4)$$

$$\Omega_g = N\Omega_r, \quad (5)$$

where J is the inertia, Ω is the rotation speed, T is the torque, k_s is the torsional stiffness, c_s is the torsional damping, θ is the rotation angle, B is the damping, and N is the gear ratio. Moreover, a represents the aerodynamic, r and g mean the rotor and the generator, respectively.

The pitch actuator and the generator were modeled as 1st-order systems, which are shown in Equations (6) and (7), respectively [25].

$$\frac{\beta}{\beta_{cmd}} = \frac{1}{1+\tau_\beta s}, \qquad (6)$$

$$\frac{T_g}{T_{g,cmd}} = \frac{1}{1+\tau_g s}. \qquad (7)$$

In Equations (6) and (7), β is the pitch angle, β_{cmd} is the pitch angle command, and τ_β is the time constant of the pitch actuator. T_g is the generator torque, $T_{g,cmd}$ is the generator torque command, and τ_g is the time constant of the generator torque.

2.4.2. The Wind Turbine Control System

Figure 7 shows the wind turbine controller used in the dynamic wind turbine model. It consists of a basic torque and blade pitch control for power maximization and regulation and a peak shaving to use blade pitching to reduce thrust force by sacrificing power slightly near the rated wind speed region [26].

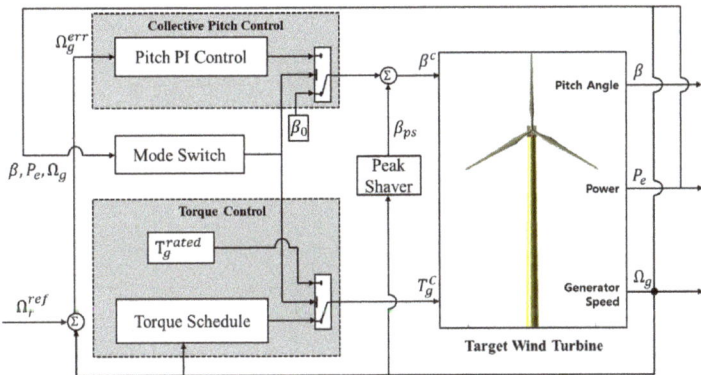

Figure 7. The wind turbine model control system.

To maximize the electrical power of the wind turbine in regions where the wind speed is lower than the rated wind speed, the pitch angle of the blade is fixed to be the fine pitch angle, and the optimal torque command is sent to the generator to maximize the power coefficient of the rotor [27]. To maintain the rated electrical power of the wind turbine in regions where the wind speed is higher than the rated wind speed, the torque is fixed to be the rated torque, and the blade pitch angle of the wind turbine is adjusted to keep the rated rotational speed of the rotor. The torque control algorithm in the control system used was a simple open-loop control using a look-up table with an input of measured generator speed and output of optimal generator torque. Moreover, the blade pitch control in the control system used the classical proportional–integral (PI) algorithm to minimize the error between the measured generator speed and the rated value.

2.5. Conversion of Power Curve to Include Air Density Effect

The power coefficient curve and thrust coefficient curve used for the wind turbine dynamic model are derived from Bladed under the condition of a standard air density of 1.225 kg/m³. However, the actual air density of a wind farm is different from the standard value, which causes a discrepancy between the prediction and the measured power output.

In this study, a novel method is proposed to match the power of a dynamic wind turbine model with that of a wind turbine with actual air density by tuning the controller. The first step of this method is to convert the power curve of the wind turbine based on the standard air density to that based on the actual mean air density of the site. This conversion is possible by using the correction method for wind turbines with active power control described in the standard, IEC 61400-12-1 [16].

Equation (8) represents the wind speed correction equation calculated based on the IEC standard.

$$V_{10min} = V_n / \left(\frac{\rho_{10min}}{\rho_0}\right)^{1/3}. \tag{8}$$

In Equation (8), V_{10min} is the measured wind speed averaged over 10 min, V_n is the normalized wind speed, ρ_{10min} is the derived 10-min average air density, and ρ_0 is the reference air density, known to be 1.225 kg/m^3. Through Equation (8), the wind speed of a power curve is converted to consider the mean air density at the site.

Figure 8 shows the comparison between the power curve with the standard air density and the power curve with the actual mean air density. It can be seen from the figure that when the power is lower than the rated power, the power with the standard air density is slightly higher than that with the actual mean air density.

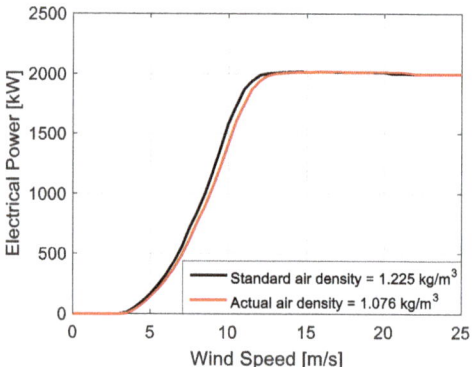

Figure 8. Comparison of the power curve of standard air density and the power curve of actual air density.

2.6. Tuning of the Wind Turbine Controller to Consider Actual Mean Air Density

To consider the actual mean air density of the site in the dynamic model, the wind turbine controller needs to be tuned by considering the power curve with the actual mean air density as a target.

In the torque control region where the wind speed is lower than 10 m/s, the look-up table of the torque scheduling was tuned so that the power curve of the dynamic model is matched with the power curve with the actual mean air density. In the transition region where the wind speed is higher than 10 m/s and lower than the rated wind speed, the peak shaving algorithm and torque scheduling were used to match the target power curve. For the rated power region, the closed-loop PI control strategy was used, and no tuning was necessary because the algorithm is not affected by the change of air density.

Figure 9 compares the target power curve with the power curve obtained from the dynamic model after tuning the control algorithms. The two power curves are well matched.

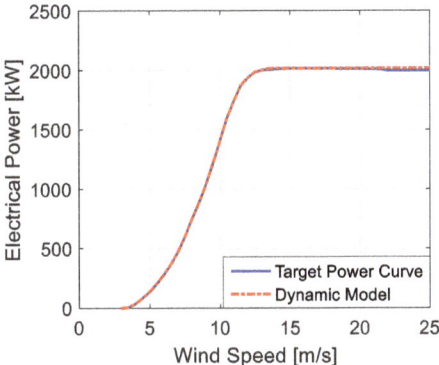

Figure 9. Comparison of the measured power curve report and the power curve of different controllers at 1.076 kg/m^3.

3. Results

Figure 10a–d shows the results of the comparison between the measured electrical power, the electrical power predicted by WindSim, and the electrical power predicted by the dynamic model for four days in January, April, August, and November, respectively. The wind direction data measured from the met mast are also shown in the figure. The measured data and the data obtained from WindSim were 10-min averaged data, and therefore, the data from the dynamic model, which was originally in 1-s intervals, were converted to 10 min averaged data.

As shown in the figure, both predictions by WindSim and the dynamic model were close to the measured power; however, there are regions where relatively large discrepancies exist in Figure 10b,d. For these regions, the wind directions are between 110 degrees and 260 degrees. The reason for these relatively large discrepancies in these regions is that the met mast is located within the wake region of the wind turbines around the target wind turbine for those wind directions, and the wind speed implemented with WindSim to predict the wind speed for the target wind turbine is the wind speed reduced by the wake effect. To obtain the wind speed for the target wind turbine, WindSim again applies a wake model in the case where the wind direction corresponds to the direction causing the wake effect on the target wind turbine. Therefore, the wind speed obtained for the target wind turbine from WindSim in those wind directions becomes lower than the actual wind speed of the target wind turbine and the power output predicted by WindSim and the dynamic model becomes lower than the measured power.

To compare the results in more detail, small portions of Figure 10c,d corresponding to three hours are replotted in Figures 11 and 12, respectively. Figure 11 shows a mean wind speed lower than the rated wind speed, and Figure 12 shows a mean wind speed higher than the rated value. Figures 11a and 12a show the wind speed measured from the met mast and the wind speed of the target wind turbine predicted by WindSim. For both cases, the wind speed of the target wind turbine predicted by WindSim is slightly higher than the wind speed measured from a nearby met mast. This is partly due to the fact that the altitude of the target wind turbine location is slightly higher than that of the met mast location.

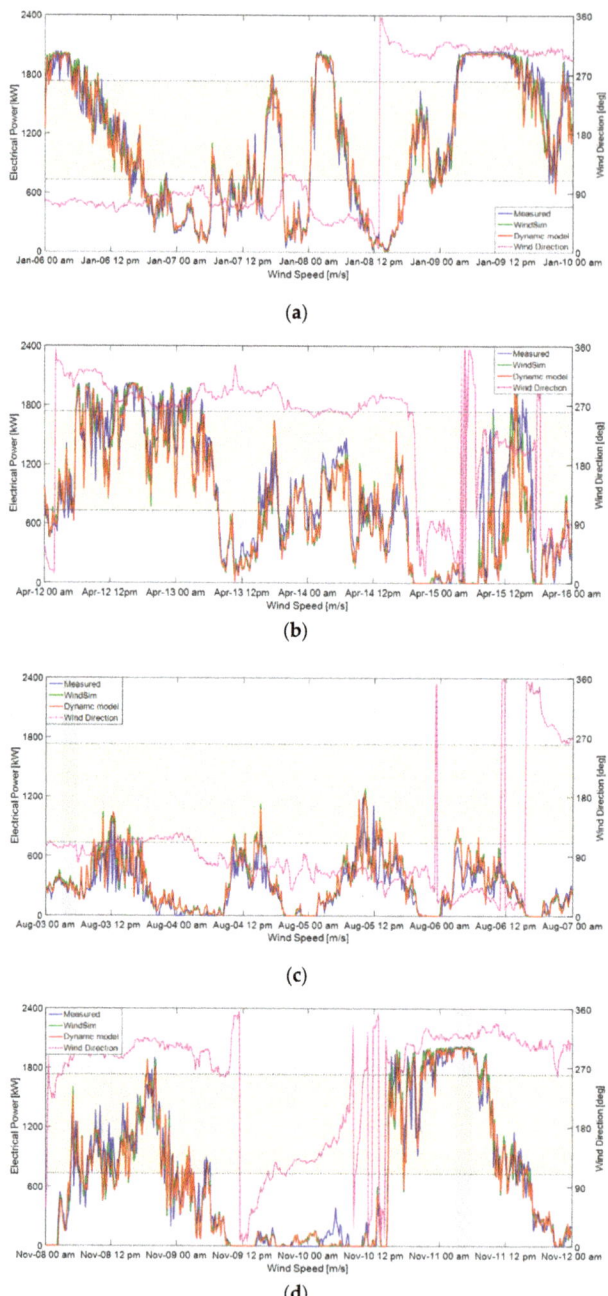

Figure 10. Comparison of the measured and predicted electrical power: (**a**) January, (**b**) April, (**c**) August, (**d**) November.

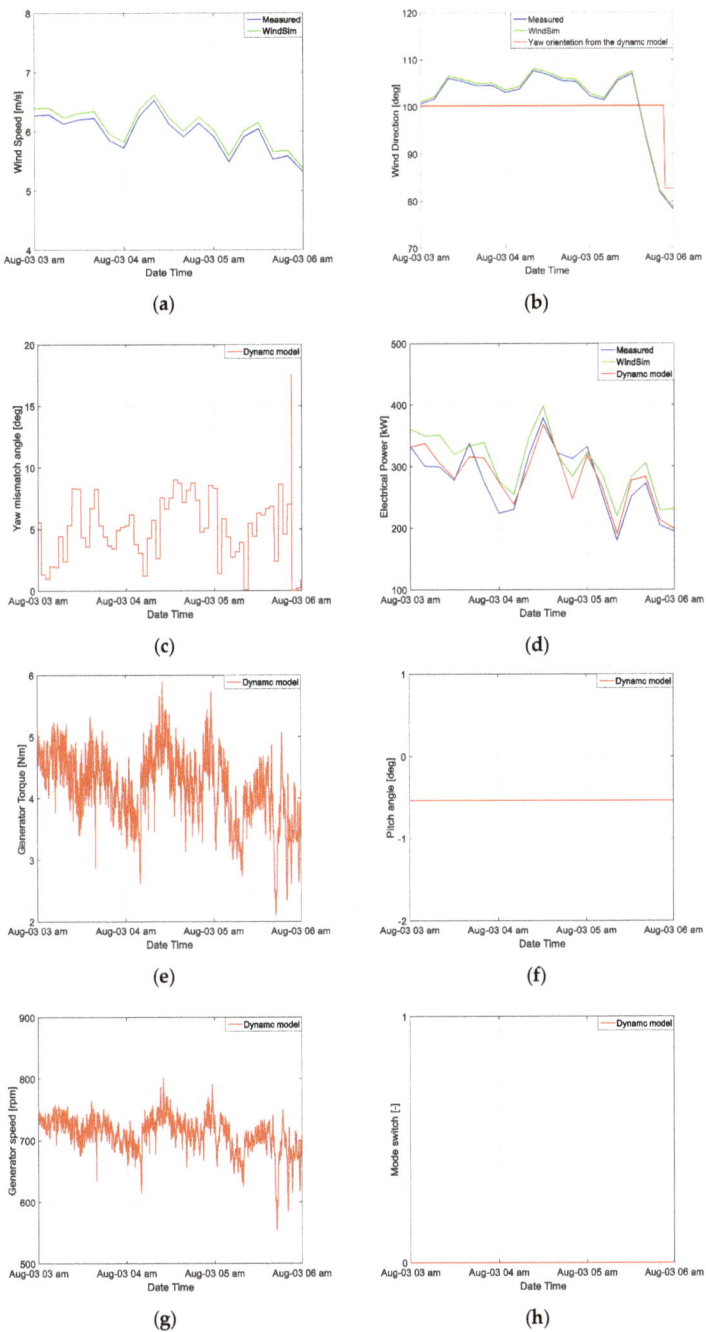

Figure 11. Comparison of the measured and predicted data for 3 h in August: (**a**) wind speed, (**b**) wind direction, (**c**) yaw mismatch, (**d**) electrical power, (**e**) generator torque, (**f**) pitch angle, (**g**) generator speed, (**h**) model switch.

Figures 11b and 12b show the wind direction measured from the met mast and the wind direction at the wind turbine position predicted by WindSim. The wind direction of the target wind turbine predicted by WindSim is basically consistent with the measured wind direction. They also show the turbine orientation predicted by the dynamic model. In Figure 11b, the yaw mismatch angle exceeded 11 degrees in the end, and the yaw system was activated to reduce the mismatch angle. In Figure 12b, the yaw orientation did not change because of the small yaw mismatch angles compared with the threshold. These are also shown in Figures 11c and 12c, which show the yaw mismatch angles. The mismatch angles are three-minute averaged values.

Figures 11d and 12d show the measured electrical power, the electrical power predicted by WindSim, and the electrical power predicted by the dynamic model. Figure 11d shows the electrical power when the wind speed is lower than the rated wind speed. Therefore, the blade pitch angle is fixed to be the fine pitch, and the generator torque is adjusted to maximize the power coefficient. As shown in the figure, the power prediction from WindSim consistently overestimates the measured power. The power predicted by the dynamic model looks similar to the WindSim results but closer to the measured power. The generator torque and the blade pitch angle from the dynamic model is shown in Figure 11e,f. Figure 12d shows the electrical power when the wind speed is mostly higher than the rated wind speed. Therefore, the generator torque is fixed to the rated value, and the blade pitch angle is adjusted to maintain the rated wind speed and the rated power. Figure 12e,f shows the generator torque and the blade pitch angle from the dynamic model. In Figure 12d, there are a few downward spikes in the measured power. The dynamic model shows similar downward spikes but the power from WindSim does not. The reason for this is that WindSim does not consider wind turbine dynamics and yaw mismatch, and the power is solely dependent on the wind speed. Based on Figure 12a, the wind speed is maintained to be mostly higher than the rated wind speed, and therefore, the power prediction from WindSim is mostly close to the rated power. However, if we look at the results from the dynamic model in Figure 12e,h, the mode switch was turned off a few times between 4:00 and 5:00 o'clock, and the generator torque was reduced. As a result, the power decreased as a downward spike. The dynamic model includes wind turbine dynamics and could show similar measured power variations.

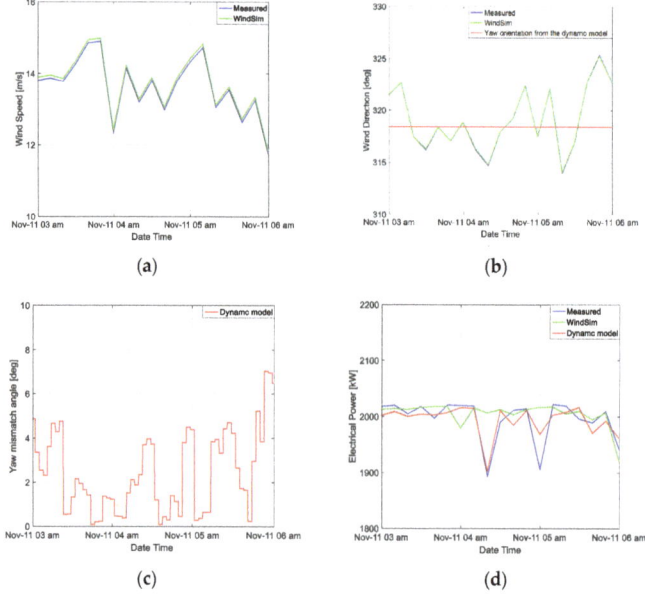

Figure 12. *Cont.*

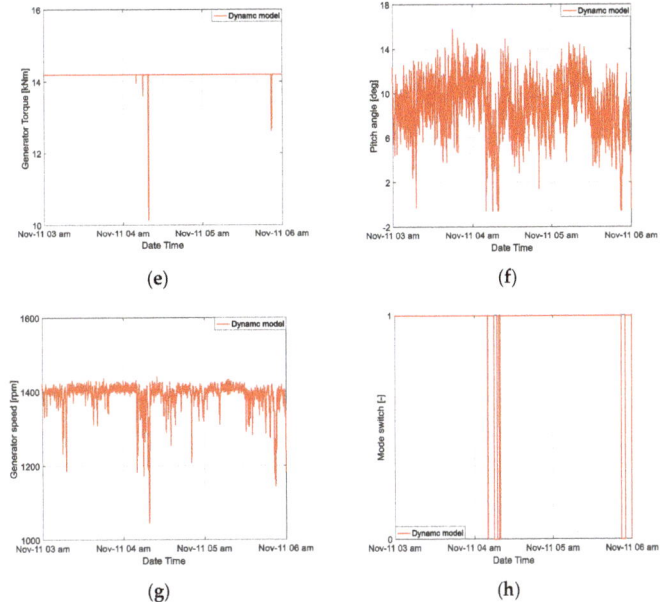

Figure 12. Comparison of the measured and predicted data for 3 h in November (**a**) wind speed, (**b**) wind direction, (**c**) yaw mismatch, (**d**) electrical power, (**e**) generator torque, (**f**) pitch angle, (**g**) generator speed, (**h**) model switch.

Figures 11g and 12g show the generator speed obtained from the dynamic model. For Figure 11g, the generator speed varies between 550 rpm and 800 rpm to maximize the power coefficient; on the other hand, for Figure 12g, the generator speed is generally close to the rated value, which is about 1400 rpm.

Table 2 shows the relative percentage errors of the electrical energy production with respect to the measure values given in Figures 11d and 12d. As shown in the table, the relative errors of the energy production predicted by WindSim and the dynamic model for Figure 11d are 9.45% and 1.12%, respectively, which is the case when the wind speed is lower than the rated wind speed. The errors are reduced to 0.48% and −0.06%, respectively, for Figure 12d in which the wind speed is higher than the rated wind speed. For both cases, the errors from the dynamic model are smaller than those from WindSim, because the dynamic model considers the wind turbine dynamics and the yaw motion. In Table 2, the electrical energy production was normalized by the measured electrical energy production (MWh) of the corresponding period, respectively.

Table 2. Relative errors of electrical energy production prediction for the data shown in Figures 10d and 11d.

	Normalized Measured	Normalized WindSim	Normalized Dynamic Model
Electrical energy production for Figure 10d	1.0000	1.0945	1.0112
Electrical energy production for Figure 11d	1.0000	1.0048	0.9994
Relative errors for Figure 10d	-	9.45%	1.12%
Relative errors for Figure 11d	-	0.48%	−0.06%

Table 3 shows the comparison between the measured AEP, the AEP predicted by the WindSim, and the AEP predicted by the dynamic model with and without yaw motion for the target wind turbine. As shown in the table, the relative error of AEP predicted by WindSim is 2.75%, and those of the dynamic model without and with the yaw system are 2.11% and 0.16%, respectively. Based on the results, the prediction error from ignoring yaw mismatch errors is larger than the prediction error from ignoring wind turbine dynamics. The annual energy production from the dynamic model is also about 2.52% smaller than that from WindSim. This result is similar to the result in Ref. [15], where the relative differences between WindPRO and the dynamic model for four different sites with the NREL 5 MW wind turbine were between 0.83% and 4.91%. In Table 3, the AEP is normalized by the measured AEP (MWh) for one year.

Table 3. The relative errors of the annual energy production (AEP) by the measured and other models.

	Normalized Measured	Normalized WindSim	Normalized Dynamic Model without Yaw	Normalized Dynamic Model with Yaw
AEP	1.0000	1.0275	1.0211	1.0016
Relative error	-	2.75%	2.11%	0.16%

4. Conclusions

In this study, a dynamic wind turbine model was constructed for an actual 2 MW wind turbine, and the wind turbine performance was predicted and validated experimentally. The dynamic model was constructed using Matlab/Simulink to consider the wind turbine dynamics, including yaw systems and a wind turbine controller using basic torque and pitch control algorithms with peak shaving in the transition from maximum power point tracking to the rated power regions. In addition, to consider the density effect of the test site, the torque scheduling and the peak shaving were tuned to reproduce the power curve with the actual mean air density.

For experimental validation, the AEP predicted by the dynamic model was compared with that by WindSim and the measured power. As the result, the relative error of WindSim was 2.75%, and those of the dynamic model without and with yaw were 2.11% and 0.16%, respectively. Although the results are only for a single wind turbine case, the prediction of AEP from the dynamic model was found to be accurate. However, further research is needed for different wind turbines and sites to better understand the performance of the dynamic simulation model.

Author Contributions: Y.S. performed the simulation of the wind turbine, analyzed the results, and wrote the paper. I.P. supervised the research, analyzed the results, and revised the paper. All authors have read and agreed to the published version of the manuscript.

Funding: This work was supported by the Human Resources Program in Energy Technology and the New and Renewable Energy-Core Technology Program of the Korea Institute of Energy Technology Evaluation and Planning (KETEP) with a granted financial resource from the Ministry of Trade, Industry and Energy, Republic of Korea (Grants No. 20173010025010 and 20204030200010).

Acknowledgments: The authors would like to acknowledge the support from Han Jin Ind. Co. Ltd. for the wind turbine model construction. The authors would also like to acknowledge the support from Jeju Energy Corp. for the climatology data of the site.

Conflicts of Interest: The authors declare no conflict of interest.

References

1. "Renewable energy 3020" Implementation Plan. Available online: http://www.motie.go.kr/motiee/presse/press2/bbs/bbsView.do?bbs_seq_n=159996&bbs_cd_n=81 (accessed on 3 August 2020). (In Korean)
2. South Korea Wind Turbine Installation Status. Available online: http://kweia.or.kr/bbs/board.php?bo_table=sub03_03 (accessed on 3 August 2020). (In Korean).
3. Yue, C.; Liu, C.; Tu, C.; Lin, T. Prediction of Power Generation by Offshore Wind Farms Using Multiple Data Sources. *Energies* **2019**, *12*, 700. [CrossRef]
4. Song, Y.; Kwon, I.; Paek, I. Investigation on promising offshore wind farm sites and their wind farm capacity factors considering wake losses in Korea. *J. Wind Energy* **2018**, *9*, 27–36. [CrossRef]
5. Park, U.; Yoo, N.; Kim, J.; Kim, K.; Min, D.; Lee, S.; Paek, I.; Kim, H. The selection of promising wind farm sites in Gangwon province using multi exclusion analysis. *J. Korean Sol. Energy Soc.* **2015**, *35*, 1–10. [CrossRef]

6. Dhunny, A.; Lollchund, M.; Rughooputh, S. Wind energy evaluation for a highly complex terrain using Computational Fluid Dynamics (CFD). *Renew. Energy* **2017**, *101*, 1–9. [CrossRef]
7. Song, Y.; Kim, H.; Byeon, J.; Paek, I.; Yoo, N. A Feasibility Study on Annual Energy Production of the Offshore Wind Farm using MERRA Reanalysis Data. *J. Korean Sol. Energy Soc.* **2015**, *35*, 33–41. [CrossRef]
8. Kim, J.; Kwon, I.; Park, U.; Paek, I.; Yoo, N. Prediction of Annual Energy Production of Wind Farms in Complex Terrain using MERRA Reanalysis Data. *J. Korean Sol. Energy Soc.* **2014**, *34*, 82–90. [CrossRef]
9. Tabas, D.; Fang, J.; Porté-Agel, F. Wind Energy Prediction in Highly Complex Terrain by Computational Fluid Dynamics. *Energies* **2019**, *12*, 1311. [CrossRef]
10. Kim, H.; Song, Y.; Paek, I. Predicted and Validation of Annual Energy Production of Garyeok-do Wind Farm in Saemangeu Area. *J. Wind Energy* **2018**, *9*, 19–24.
11. Kim, H.; Kim, K.; Paek, I. Power regulation of upstream wind turbines for power increase in a wind farm. *Int. J. Precis. Eng. Manuf.* **2016**, *17*, 665–670. [CrossRef]
12. Kim, H.; Kim, K.; Paek, I. Model Based Open-Loop Wind Farm Control Using Active Power for Power Increase and Load Reduction. *Appl. Sci.* **2017**, *7*, 1068. [CrossRef]
13. Kim, H.; Kim, K.; Paek, I. A Study on the Effect of Closed-Loop Wind Farm Control on Power and Tower Load in Derating the TSO Command Condition. *Energies* **2019**, *12*, 2004. [CrossRef]
14. Campagnolo, F.; Weber, R.; Schreiber, J.; Bottasso, C. Wind tunnel testing of wake steering with dynamic wind direction changes. *Wind Energy Sci.* **2020**, *5*, 1273–1295. [CrossRef]
15. Kim, K.; Lim, C.; Oh, Y.; Kwon, I.; Yoo, N.; Paek, I. Time-domain dynamic simulation of a wind turbine including yaw motion for power prediction. *Int. J. Precis. Eng. Manuf.* **2014**, *15*, 2199–2203. [CrossRef]
16. International Electrotechnical Commission (IEC). *Wind Turbine Generator Systems Part 12-1: Power Performance Measurements of Electricity Producing Wind Turbines*, 2nd ed.; International Electrotechnical Commission (IEC): Geneva, Switzerland, 2017.
17. Ministry of Land. Infrastructure and Transport, National Spatial Data Infrastructure Portal. Available online: http://www.nsdi.go.kr/lxportal/?menuno=2679 (accessed on 3 August 2020).
18. Ministry of Environment. Environmental Spatial Information Service. Available online: http://egis.me.go.kr/map/map.do?type=land (accessed on 3 August 2020).
19. Hwang, Y.; Paek, I.; Yoon, K.; Lee, W.; Yoo, N.; Nam, Y. Application of wind data from automated weather stations to wind resources estimation in Korea. *J. Mech. Sci. Technol.* **2010**, *24*, 2017–2023. [CrossRef]
20. WindSim AS. Available online: https://www.windsim.com (accessed on 3 August 2020).
21. Cham. The Phoenics Encyclopedia. Available online: http://www.cham.co.uk/phoenics/d_polis/d_enc/enc_gcv.htm (accessed on 3 August 2020).
22. Fallo, D. Wind Energy Resource Evaluation in a Site of Central Italy by CFD Simulations. Ph.D. Thesis, University of Cagliari, Cagliari, Italy, 2007.
23. Katic, I.; Højstrup, J.; Jensen, N. A simple model for cluster efficiency. In Proceedings of the European Wind Energy Association Conference and Exhibition, Roma, Italy, 7–9 October 1986; Volume 1, pp. 407–410.
24. Nam, Y.; Kim, J.; Paek, I.; Moon, Y.; Kim, S.; Kim, D. Feedforward Pitch Control Using Wind Speed Estimation. *J. Power Electron.* **2011**, *11*, 211–217. [CrossRef]
25. Bianchi, F.; de Battista, H.; Mantz, R. Wind Turbine Control Systems, Principles, Modelling and Gain Scheduling Design. Available online: https://www.springer.com/gp/book/9781846284922 (accessed on 3 August 2020).
26. Kim, K.; Kim, H.; Paek, I. Application and Validation of Peak Shaving to Improve Performance of a 100 kW Wind Turbine. *Int. J. Precis. Eng. Manuf.-Green Tech.* **2020**, *7*, 411–421. [CrossRef]
27. Jonkman, J.; Butterfield, S.; Musial, W.; Scott, G. *Definition of a 5-MW Reference Wind Turbine for Offshore System Development*; National Renewable Energy Lab. (NREL): Golden, CO, USA, 2009; p. 947422.

Publisher's Note: MDPI stays neutral with regard to jurisdictional claims in published maps and institutional affiliations.

© 2020 by the authors. Licensee MDPI, Basel, Switzerland. This article is an open access article distributed under the terms and conditions of the Creative Commons Attribution (CC BY) license (http://creativecommons.org/licenses/by/4.0/).

Article

Geometry Design Optimization of a Wind Turbine Blade Considering Effects on Aerodynamic Performance by Linearization

Kyoungboo Yang

Faculty of Wind Energy Engineering, Jeju National University, 102 Jejudaehakno, Jeju 63243, Korea; kbyang@jejunu.ac.kr; Tel.: +82-64-754-4405

Received: 3 April 2020; Accepted: 4 May 2020; Published: 7 May 2020

Abstract: For a wind turbine to extract as much energy as possible from the wind, blade geometry optimization to maximize the aerodynamic performance is important. Blade design optimization includes linearizing the blade chord and twist distribution for practical manufacturing. As blade linearization changes the blade geometry, it also affects the aerodynamic performance and load characteristics of the wind turbine rotor. Therefore, it is necessary to understand the effects of the design parameters used in linearization. In this study, the effects of these parameters on the aerodynamic performance of a wind turbine blade were examined. In addition, an optimization algorithm for linearization and an objective function that applies multiple tip speed ratios to optimize the aerodynamic efficiency were developed. The analysis revealed that increasing the chord length and chord profile slope improves the aerodynamic efficiency at low wind speeds but lowers it at high wind speeds, and that the twist profile mainly affects the behaviour at low wind speeds, while its effect on the aerodynamic performance at high wind speeds is not significant. When the blade geometry was optimized by applying the linearization parameter ranges obtained from the analysis, blade geometry with improved aerodynamic efficiency at all wind speeds below the rated wind speed was derived.

Keywords: design optimization; wind turbine blade; blade geometry linearization; tip speed ratio; simulated annealing algorithm

1. Introduction

The purpose of a wind turbine is to produce a large amount of electrical energy by extracting as much energy as possible from the wind. To this end, the efficiency of each component is optimized in the wind turbine design procedure. In this design procedure, the wind turbine blade geometry is optimized first to maximize the aerodynamic performance of the turbine [1]. The optimized blade shape allows greater use of wind energy, and the aerodynamic performance of a wind turbine is an important factor in its design. When a wind turbine is designed, however, its weight, load, and manufacturability must also be considered. Consequently, wind turbine designers have attempted to find blade geometry optimization methods to achieve maximum power and minimum manufacturing cost. Although proper compromises have been found on practical levels, research on the best blade geometry design method is still required [2].

Studies on wind turbine blade design have been continuously conducted based on the aerodynamic performance analysis theory of aircraft propellers established by Glauert [3]. Wilson et al. proposed a performance prediction method for vertical and horizontal axis wind turbines by applying the Glauert method to wind turbine blades [4,5]. Based on this work, a blade element momentum (BEM) theory for wind turbine geometry design was established, and studies on various blade design methods and performance analysis have been conducted using this theory [6–10]. Furthermore, generalized wind turbine rotor blade design procedures based on the BEM theory are now available [1,2]. However, these procedures deal with ideal blade geometry, and the ideal design obtained from the BEM theory is heavy and difficult to manufacture due to the nonlinear chord profile. Therefore, the chord and twist profile are linearized considering manufacturability and weight. In the blade linearization process, the actual aerodynamic efficiency is inevitably lower than that of the ideal blade geometry. Consequently, an optimal linearization method is required to minimize the aerodynamic efficiency reduction, and various blade chord and twist profile linearization methods have been proposed.

Manwell et al. [1] developed a linear equation with three coefficients for chord and twist profile linearization. To apply this equation in blade linearization, it is necessary to optimize three coefficients for maximum efficiency. Burton et al. [2] used a straight line that passes through the 70% and 90% positions of the blade span for chord linearization. Maalawi and Badr [11] mentioned that the chord profile must form a tangent at the 75% position of the theoretical blade span and proposed an exponential distribution equation for the twist profile. Liu et al. [12] proposed a linear equation that performs linearization based on the theoretical blade tip chord and twist profile for fixed-pitch, fixed-speed wind turbine blades. Yang et al. [13] linearized the blade chord and twist profile using a Bezier curve that adjusts five control points. Bezier curves are parametric curves that are mainly used for geometric design of products requiring smooth shapes, and the curve geometry is determined by the positions of the control points. Therefore, the coordinates of the control points that maximize the aerodynamic efficiency must be found. Tahani et al. [14] used a method of selecting a line with the highest efficiency among all lines that connect two adjacent points along the theoretical blade span, but this method requires many blade element sections to increase the precision. They performed division into 30 blade section elements. Most of the linearization methods used in previous studies require several parameters for linearization, even though there are also simple methods; thus, it is necessary to find the optimal combination of such parameters.

Most linearization methods are focused on maximizing the aerodynamic efficiency based on the specific design wind speed and design tip speed ratio (TSR). However, blades designed based on particular design wind speeds and design TSRs exhibit maximum efficiency at specific wind speeds [15]. Wind turbines perform control to track the maximum power point, but the ability to respond to the wind, which varies every moment, is limited. Therefore, blade geometry optimization that considers multiple TSRs is required to achieve optimal efficiency at various wind speeds. In addition, as the blade linearization process affects the aerodynamic performance and blade stiffness due to the geometry change, blade design optimization becomes very complicated if the structural stability of the blade is also considered. Therefore, blade geometry optimization, including linearization, must deal with various constraints. To achieve the performance goals of wind turbines, optimization algorithms have been applied to blade geometry design in various studies.

Selig and Coverstone-Carroll [16] performed blade geometry optimization using a genetic algorithm (GA) in which the annual energy production was maximized as an objective function. The blade chord and twist profile linearization was determined based on the 75% position of the theoretical blade span. Fuglsang and Madsen [17] optimized the geometry of a 1.5 MW stall-regulated rotor blade using sequential linear programming and the method of feasible directions. Jureczko et al. [18] performed multi-criterion design optimization using a GA considering the structural conditions, including the aerodynamic load and material of the wind turbine blade. In this research, a finite element model was used, and the focus was on the structural performance of the blade rather than on aerodynamic performance. Mendez and Greiner [19] optimized the blade geometry to maximize the average power dependent on the Weibull wind distribution at a specific site using a GA. Similarly, Polat and Tuner [20] optimized the aerodynamic shape of wind turbine blades using a parallel GA. Further, Tahani et al. [21] determined the blade chord and twist profile as well as the optimal blade section positions for various airfoil types using a GA. Ashuri et al. [22] presented a method for multidisciplinary design optimization at the system level considering integrated aerodynamic and structural design of the rotor and tower simultaneously. Subsequently, Neto et al. [23] performed blade geometry optimization using an evolutionary algorithm to maximize the energy production of wind turbines and to minimize the mass of the blade. In most studies on blade geometry optimization, the rotor blade geometry has been optimized to maximize the wind turbine power or energy. As the blade geometry is determined by the blade chord and twist profiles, the chord and twist linearization, which determine the blade geometry, play important roles in blade geometry optimization.

This study deals with the chord and twist linearization and geometry optimization required for blade geometry design. As the blade linearization process changes the blade geometry, it also affects the aerodynamic performance and load characteristics of the wind turbine rotor. Therefore, it is necessary to understand the effects of the parameters that are used in the linearization process. Some of the abovementioned studies dealt with blade linearization, but there has been no discussion of the effects of the linearization-induced changes in blade geometry on the aerodynamic performance. In this study, the effects of the wind turbine blade linearization parameters on the aerodynamic performance of the wind turbine blade were examined. In addition, an optimization algorithm for linearization and an objective function that applies multiple TSRs to satisfy the optimal efficiency at various wind speeds were established. The linearization effect analysis and blade design optimization using the derived algorithm were conducted for a 5-MW baseline wind turbine developed by the National Renewable Energy Laboratory (NREL), and its aerodynamic performance was verified using FAST (Fatigue, Aerodynamics, Structures and Turbulence) code. In addition, the aerodynamic performance of the blade optimized using the developed algorithm was compared with that of the NREL baseline blade to verify the performance of the proposed optimization algorithm and the suitability of the objective function.

2. Blade Design Procedure and Methodology

2.1. Blade Design Procedure

Among the methodologies for wind turbine blade design and analysis, the BEM theory has been widely used due to its short calculation time and satisfactory results [24]. This theory combines the blade elements with one-dimensional momentum theory, as detailed in various literature [1,2]. This section presents the BEM theory equations that were used in the optimal blade design procedure in this study, which are summarized in Figure 1. In addition, it describes the overall flow, including the basic blade geometry design, optimization process for chord and twist angle linearization, and aerodynamic performance analysis process.

2.2. Initial Blade Geometry Design

The initial blade geometry design is obtained by referring to the theoretical or ideal geometry and begins with the selection and placement of airfoils along the span of the blade. Initially, the width of all airfoils is set to the same, and then the appropriate airfoil position is determined through optimization. In this study, the selection of airfoils and their positioning are based on the section layout of the NERL baseline blade. After sections are divided along the span of the blade and airfoils are selected, the angle of attack (AoA, α) of each airfoil is determined from the corresponding airfoil data. In addition, the local TSR (λ_r) is calculated according to the rotor radial position (r) and ideal inflow angle (φ_r) of the wind for each section (see Figure 2), as follows:

$$\varphi_r = \tan^{-1}\left(\frac{2}{3\lambda_r}\right). \tag{1}$$

Equation (1) assumes $a = 1/3$ and $a' = 0$ in Equation (7), i.e., the Betz limit and no wake rotation. Once the ideal inflow angle has been calculated, the chord length (c_r) and twist angle (θ_r) of each blade section can be calculated as follows:

$$c_r = \frac{8\pi r}{Nc_L}\left(\frac{\sin\varphi}{3\lambda_r}\right), \tag{2}$$

$$\theta_r = \varphi_r - \alpha, \tag{3}$$

where N is the number of blades and c_L is the lift coefficient to the AoA of the airfoil at section r.

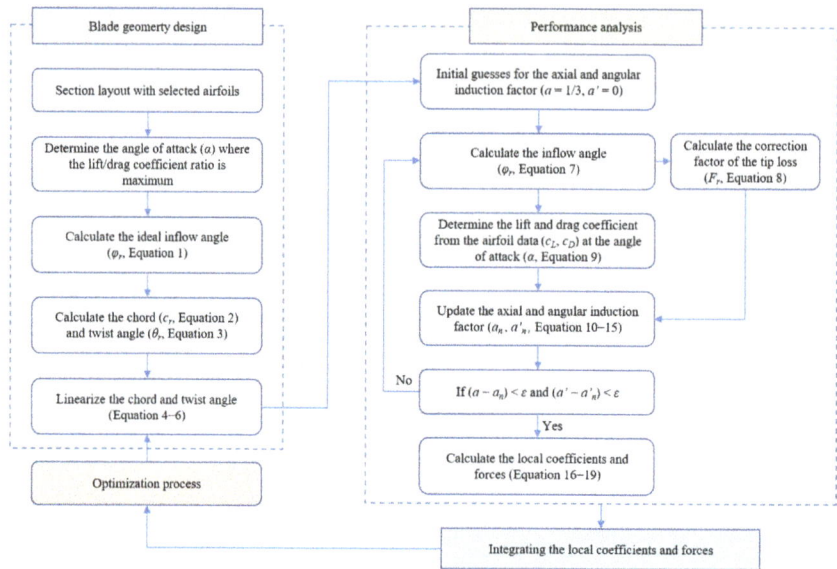

Figure 1. Optimal blade design procedure using BEM theory and optimization algorithm.

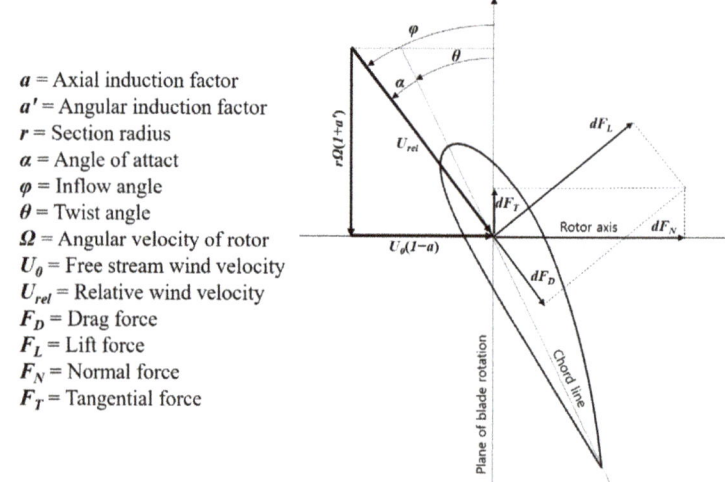

a = Axial induction factor
a' = Angular induction factor
r = Section radius
α = Angle of attact
φ = Inflow angle
θ = Twist angle
Ω = Angular velocity of rotor
U_0 = Free stream wind velocity
U_{rel} = Relative wind velocity
F_D = Drag force
F_L = Lift force
F_N = Normal force
F_T = Tangential force

Figure 2. Wind velocities and forces on the airfoil in a blade section [1].

Using the above process, theoretical blade geometry in which the chord length increases toward the blade root section is created, and linearization is performed on the ideal chord and twist distributions. Various linearization methods have been proposed, as mentioned in Section 1, but the linear method with the most common linear slope [1] was used in this study. The linearization equations for the chord and twist distribution are as follows:

$$C_r = a_1 r + b_1, \quad (4)$$

$$\theta_r = a_2(R - r), \quad (5)$$

where R is the radius of the wind turbine rotor, a_1 is the chord profile slope and a_2 is the coefficient defining the twist distribution of the blade. b_1 is an intercept of the chord linear function and can be calculated using the chord length (TC) at the reference radius (r_{TC}) as follows:

$$b_1 = TC - a_1 r_{TC}. \quad (6)$$

2.3. Blade Performance Analysis

After performing chord and twist profile linearization, the blade geometry is practical. Performance analysis of this blade is also conducted using the BEM theory. The core of the blade performance analysis is obtaining the practical axial and angular induction factors (a, a') of the designed blade. As a and a' cannot be calculated analytically, the inflow angle of the wind and the corresponding AoA are calculated first by estimating a and a', and the induction factor that satisfies the designed blade is obtained while repeating the process of deriving new a_n and a'_n, as shown in Figure 1. For the initial estimate of the induction factors, $a = 1/3$ and $a' = 0$ are generally applied. The inflow angle can be calculated using the axial and angular induction factors as follows:

$$\varphi_r = \tan^{-1}\left(\frac{1-a}{(1+a')\lambda_r}\right). \quad (7)$$

The correction factor of the tip loss to reflect the lift force reduction due to airflow around the blade tip is based on Prandtl's method as follows:

$$F_r = \left(\frac{2}{\pi}\right)\cos^{-1}\left[\exp\left(-\left\{\frac{N(R-r)}{2r\sin\varphi_r}\right\}\right)\right]. \quad (8)$$

The AoA of the airfoil is calculated using the inflow angle (Equation (7)) and the twist angle derived from the linearization process above, and c_L and c_D are determined from the lift and drag data of the airfoil.

$$\alpha = \varphi_r - \theta_r, \quad (9)$$

New a_n and a'_n values are calculated using the calculated inflow angle and correction factor as follows:

$$a_n = \frac{1}{\left(1 + \frac{4F_r \sin^2 \varphi_r}{\sigma_r C_N}\right)}, \quad (10)$$

$$a'_n = \frac{1}{\frac{4F_r \sin\varphi_r \cos\varphi_r}{\sigma_r C_T} - 1}, \quad (11)$$

where σ_r is the local solidity ($\sigma_r = Nc_r/2\pi r$), C_N is the normal force coefficient and C_T is the tangential force coefficient. These quantities can be calculated using the c_L and c_D values of the airfoil, determined above as follows:

$$C_N = c_L \cos\varphi_r + c_D \sin\varphi_r, \quad (12)$$

$$C_T = c_L \sin\varphi_r - c_D \cos\varphi_r. \quad (13)$$

Once new induction factors (a_n, a'_n) have been calculated, the differences from the induction factors (a, a') estimated above are examined, and this process is repeated until the error range (ε)

desired by the designer is achieved. During this process, the new axial induction factor (a_n) may exceed the Betz limit. As this situation causes a conflict with the BEM theory, the following correction is performed:

$$a_n = \begin{cases} \text{Equation (10)}, \ a > 0.2 \\ \frac{1}{2}\left[2 + K(1-2a) - \sqrt{(K(1-2a)+2)^2 + 4(Ka^2 - 1)}\right] \end{cases}, \quad (14)$$

$$K = \frac{4F_r \sin^2 \varphi_r}{\sigma_r C_N}. \quad (15)$$

When satisfactory induction factors (a, a') are derived at a given section position, the local power coefficient ($C_{p,r}$) and thrust coefficient ($C_{t,r}$) can be calculated as follows:

$$C_{p,r} = \frac{8}{\lambda^2} F_r \lambda_r^3 a'(1-a)[1 - (c_D/c_L)\cot\theta_r], \quad (16)$$

$$C_{t,r} = \sigma_r(1-a)^2 C_N / \sin^2\varphi_r. \quad (17)$$

The normal force (dF_n) and tangential force (dF_t) acting on the blade can be calculated from the axial and angular momentum as follows:

$$dF_n = F_r \rho U^2 4a(1-a)\pi r dr, \quad (18)$$

$$dF_t = F_r \rho U 4a'(1-a)\pi r^3 \Omega dr, \quad (19)$$

where ρ is the air density, U is free stream wind speed, and Ω is the angular velocity of the rotor.

2.4. Optimization Algorithm for Linearization

Wind turbine blade geometry optimization is the process of determining the geometry that can generate the maximum aerodynamic efficiency or maximum energy of the wind turbine. From a technical perspective, it involves determining the linearized optimal chord and twist angle distribution. As shown in Figure 1, the optimization process is applied to the chord and twist angle linearization during blade design. The process of the optimization algorithm is summarized in Figure 3. The optimization algorithm used in this study was a simulated annealing (SA) algorithm. The SA algorithm is among the representative global optimization methods, along with GA, and simulates the annealing process in metallurgy [25].

The main steps in the SA algorithm are initialization, perturbation, and evaluation, and the temperature parameter controls all of these steps. The temperature parameter is not the actual temperature, but rather a computational parameter that controls the solution selection probability in the process of finding an optimal solution in the SA algorithm. As shown in Figure 3, the geometry derived from the theoretical blade design is initialized, and the perturbation process finds the coefficients of the linear functions (Equations (4), (5) and (6)) for the chord and twist distribution. A solution search is performed on a probabilistic basis and controlled by the abovementioned temperature parameter. The initial temperature parameter may be set to 1.0 or lower depending on the characteristics of the problem. The temperature parameter gradually decreases in a manner similar to that in the annealing process for metals. A linearly fast temperature decrease increases the possibility of convergence to local optima, and a logarithmically slow decrease increases the performance time. In the evaluation step, the optimization objective function is evaluated.

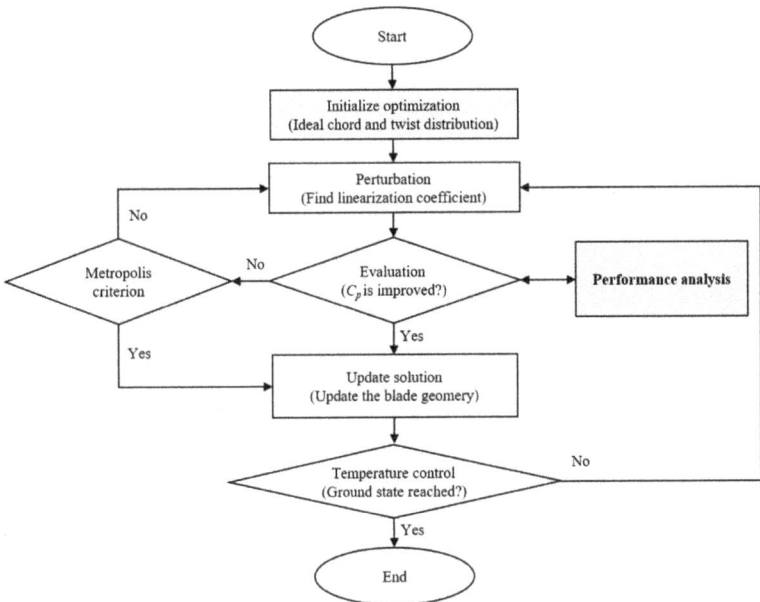

Figure 3. Optimization process for linearization of the chord and twist distributions.

The Metropolis criterion is a selection process that determines whether to accept a candidate solution or not. Basically, a candidate solution will be accepted as the current solution if the objective function value improves in the evaluation. However, even a candidate solution ($S_{candidate}$) that became worse during the perturbation, depending on the probability of the Metropolis criterion, can be accepted as the current solution ($S_{current}$). This characteristic of the SA algorithm avoids the possibility of convergence to the local optima. The Metropolis criterion, based on the Metropolis–Hastings algorithm, is as follows:

$$P_{metro} = \exp\left(-\frac{\Delta \cos t}{T}\right), \tag{20}$$

$$S_{current} = \begin{cases} S_{candidate}, & \text{if } P_{metro} > P_{rand} \\ S_{current}, & \text{otherwise} \end{cases}, \tag{21}$$

where P_{metro} is a metropolis probability, $\Delta cost$ is the difference between a candidate value and the current value, P_{rand} is a probability obtained using the random number generator function, and T is the current temperature parameter.

The parameters of the optimization algorithm were determined based on the preliminary performance. For the initial temperature, 0.3 was applied because the solutions only fluctuated without improvement until approximately 0.3. A high temperature parameter increased the probability of selecting a worse solution. For the stopping temperature, a marginal value in which there is no improvement of the solution even after a long execution time was selected. The related parameters are listed in Table 1.

Table 1. Parameters of the optimization algorithm.

Parameter	Value
Initial temperature (−)	0.3
Stopping temperature (−)	0.001
Temperature control	$T_i = 0.98 T_{i-1}$
Iteration at each temperature	200

In this study, the purpose of blade geometry optimization was to maximize the aerodynamic performance, i.e., to maximize the power coefficient. However, the maximum power coefficient varies depending on the design TSR, and a blade design based on a specific design TSR exhibits the maximum efficiency only under the specific corresponding wind conditions. As the wind varies continuously, it is necessary to consider the power coefficient at various operational TSRs to ensure satisfactory blade performance under different wind speed conditions. Therefore, in this study, the following objective function that considers multiple operational TSRs was proposed to consider off-design performance:

$$F_{obj} = \sum_{\lambda=L_1}^{L_2} \left(\int_{\lambda_h}^{\lambda} C_{p,r}(\lambda) dr_\lambda \right), \quad (22)$$

where λ_h is the local speed ratio at the hub. L_1 and L_2 are the lower and upper limits of the design TSR range. As the blade geometry and performance vary depending on the TSR design range, it is necessary to select a design range that can satisfy the target performance of the blade. Further details are provided in Section 3.

3. Effects of Linearization on Blade Performance

The effects of linearization on the aerodynamic performance of the blade were examined before optimizing the blade geometry. In addition, appropriate ranges of the coefficients of the linear functions (Equations (4)–(6)) to be applied in the blade geometry optimization were selected. If the ranges of the linearization coefficients to be searched by the algorithm are not determined, very large or thin blade geometry may exhibit high efficiency, and, thus, the optimization algorithm may derive unrealistic geometry as the final result. In addition, understanding the effects of changes in the chord and twist profiles caused by linearization on the aerodynamic performance of the blade can be helpful for blade design.

3.1. NREL 5 MW Baseline

The 5 MW baseline wind turbine developed by NREL [26] was used to evaluate the effects of blade linearization and to verify the proposed optimization algorithm. This wind turbine is a utility-scale turbine developed by NREL to support concept studies, and various research has been conducted based on the specifications of this wind turbine. Table 2 shows the specifications of the baseline wind turbine. In this study, the effects of linearization were evaluated by varying the chord and twist distribution based on the basic specifications of the NREL 5 MW baseline turbine blade, and new blade geometry was designed using the proposed optimization algorithm.

Table 2. Properties of the NREL baseline wind turbine.

Property	Value
Rated power (MW)	5
Number of blades	3
Hub height (m)	90
Rotor diameter (m)	126
Cut-in, rated, cut-out wind speed (m/s)	3, 11.4, 25
Cut-in, rated rotor speed (rpm)	6.9, 12.1
Rated tip speed (m/s)	80

3.2. Effects of Chord and Twist Linearization

The main parameters of the chord distribution that determine the overall blade geometry are the slope (a_1), and TC, which determines the intercept (b_1) of the linear function (Equation (4)). In this study, between 19% and 93% of the blade length was set as the reference radius in order not to deform the tip geometry of the baseline. In addition, the blade performance was examined according to the TC and slope at the reference radius, as shown in Figure 4.

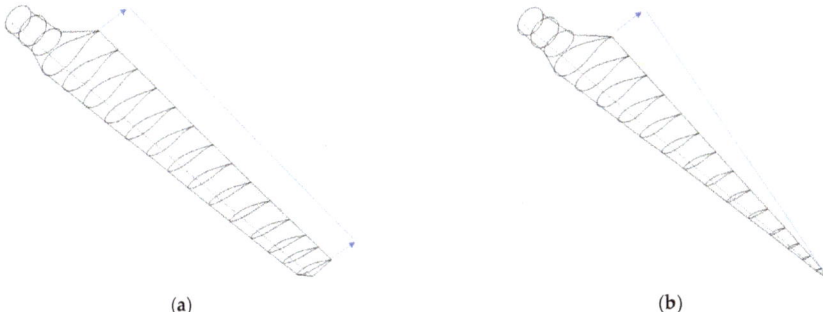

(a) (b)

Figure 4. Concept illustration for analysis of the linearization effect: (**a**) chord length and (**b**) chord profile slope.

Various chord profiles to illustrate the effects of the slope of the linear function along with a baseline chord are shown in Figure 5. The ideal chord corresponds to the theoretical chord lengths of the airfoils that constitute the baseline. The chord length significantly increases toward the root section. In addition, the ideal chord values are irregular because the baseline blade was composed of various types of airfoils instead of a single type. In the NREL baseline case, NACA-series airfoils for aerodynamic performance were used in sections close to the tip and DU-series airfoils for structural stability were used toward the root section, considering structural aspects. Figure 6 presents twist profiles to elucidate the effects of twist angle linearization on the aerodynamic performance of the blade. In the case of the twist profiles, the variation of the power coefficient (C_p) was examined while the linearization coefficient (a_2) in Equation (5) was varied. The C_p variation due to linearization was examined for each TSR. The TSR, an important parameter in blade design, varies depending on the blade design objective, and the blade geometry varies depending on the TSR. To ensure satisfactory blade performance under various wind speeds considering the off-design performance, it is necessary to understand the changes in aerodynamic performance due to linearization at various TSRs.

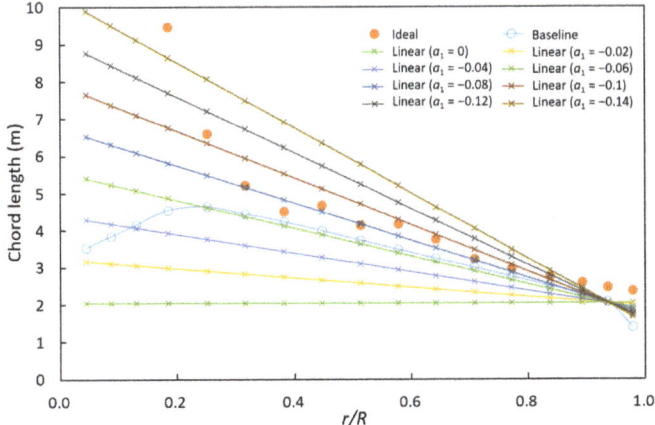

Figure 5. Chord profiles according to the slope for analysis of the linearization effect.

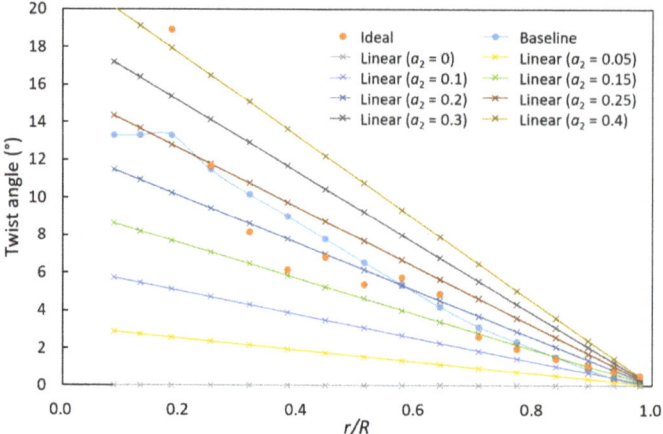

Figure 6. Twist profiles of the blade for analysis of the linearization effect.

The C_p for each TSR, according to the reference chord length (TC = 1–5 m) with the chord profile slope and twist angle profiles fixed ($a_1 = 0.6$, $a_2 = 0.2$), is shown in Figure 7. An increase in TC increases the overall blade size, as shown in Figure 4a. Further, C_p increases at low TSRs ($\lambda = 4$–6) but rapidly decreases at high TSRs ($\lambda = 9$–11) as the blade size increases. High TSRs can be seen as low wind speed conditions because the wind speed is low compared to the rotor rotation speed. Therefore, the aerodynamic performance may decrease at low wind speeds if the blade size increases. On the other hand, low TSRs correspond to high wind speed conditions, which generally correspond to the rated power region of a wind turbine. As the wind turbine power is limited through pitch control in this region, the TSR range of interest is the high TSR range. For TSR = 7–8 with the highest power coefficient, the maximum C_p is observed when the TC is approximately 3–4 m. The C_p corresponding to high TSRs, however, significantly decreases with increasing TC length. As the overall blade size increases, the blade weight will also increase. Therefore, considering the power coefficient and blade weight at high TSRs that correspond to low wind speeds, the practical design range of the TC appears to be the 1.5–2.5 m section marked in Figure 7. To design other blades, appropriate TC ranges can be set through the same analysis process.

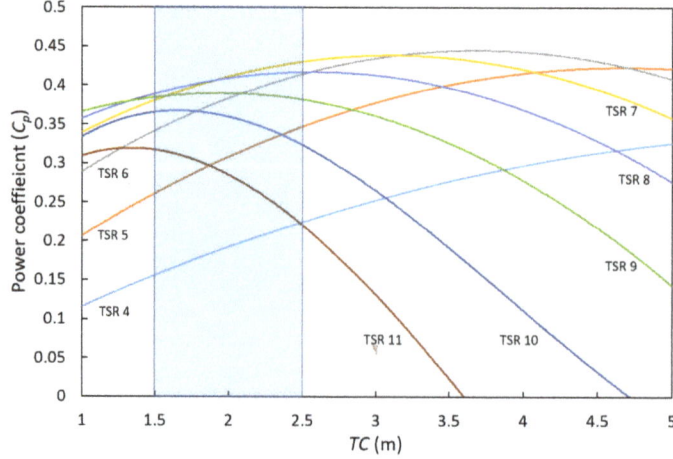

Figure 7. Variation of the power coefficient with the reference chord length at each TSR.

The C_p for each TSR according to the chord profile slope with the reference chord length and twist angle profiles fixed ($TC = 2.0$, $a_2 = 0.2$) is shown in Figure 8. The blade size increases toward the root section, as shown in Figure 4b. Similar to the C_p variation with the chord length in Figure 7, C_p increases for $TSR = 4$–6 but decreases for $TSR = 9$–11 as the slope increases. Compared to the effect of the chord length increase, however, the effect of the slope change is small. For $TSR = 7$–8, C_p is the highest at a slope of approximately $a_1 = -0.13$, but this situation is not realistic because the maximum chord length close to the root of the blade reaches 8 m. Therefore, a design range from -0.04 to -0.08 appears to be appropriate for the slope of the chord profile.

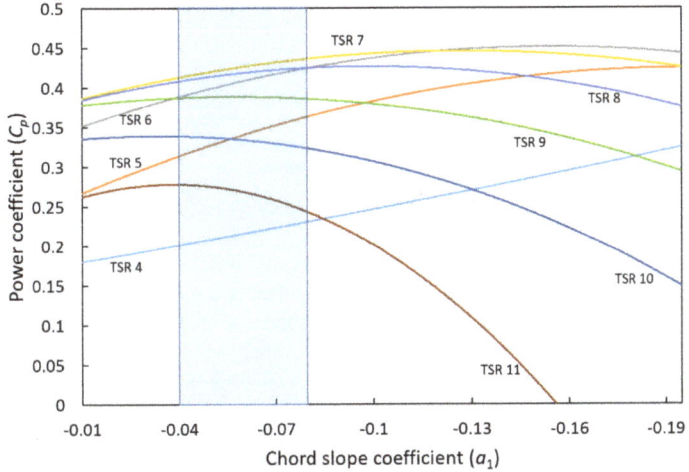

Figure 8. Variation of the power coefficient with the chord slope parameter at each TSR.

Next, the aerodynamic performance of the blade according to the twist angle profile was examined for each TSR (Figure 9). As the change in twist profile does not have large effects on the size and weight of a blade, it does not significantly limit the design range. Figure 9 shows the C_p variation patterns differing from those caused by the variations in chord length and slope described above. The C_p increases for $TSR = 4$–6, but the increment is not large. For $TSR = 9$–11, C_p rapidly increases and then sharply decreases at approximately $a_2 = 0.2$. Even for $TSR = 7$–8, where C_p reaches its maximum, C_p slowly increases and then decreases. In the case of the twist profile, the parameter range with a high power coefficient in terms of aerodynamic performance can easily be set using the graph. In this study, the range corresponding to $a_2 = 0.15$–0.25 was set as the design range, as shown in Figure 9.

Summarizing the power coefficient variation for each TSR caused by the linearization parameters, it is evident that the blade geometry has various effects on the power coefficient at each TSR. This finding indicates that the blade performance can be significantly limited if only a specific TSR is considered. Therefore, it is necessary to consider various TSRs, and one role of the objective function proposed in this report is to find the optimal combinations in various TSR ranges.

3.3. Comparison of Aerodynamic Performance

To examine the differences in the aerodynamic performance of the wind turbine blades when each blade was designed by utilizing the three linearization design ranges described above (chord length, chord slope, and twist angle profile), power coefficient was compared using FAST code developed by NREL. The comparison included the central values in the design range established in Section 3.2. Each blade was designed by applying lower and higher values with respect to the central values, and power coefficient versus TSR (C_p–λ curve) and wind speeds were compared.

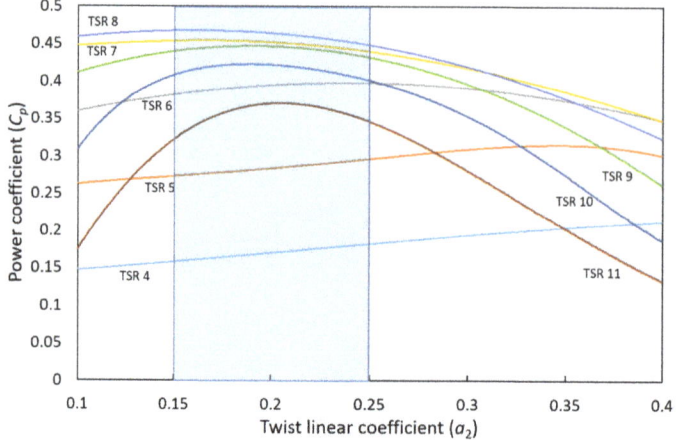

Figure 9. Variation of the power coefficient with the twist linear parameter at each TSR.

The FAST results for three blades with different reference tip chord lengths (TC = 1.5, 2, and 2.5 m) are shown in Figure 10. In Figure 10a, the C_p–λ curve is shifted to the right (high TSR) as TC decreases. The graph of C_p versus wind speed in Figure 10b reveals that the smallest TC = 1.5 m has the highest C_p, while TC = 2 and 2.5 m show lower values at low wind speeds of 3–5 m/s. For wind speeds greater than 6 m/s, however, TC = 1.5 m exhibits the lowest C_p. In other words, as the blade chord length decreases, C_p is suitable at low wind speeds but not at high wind speeds. The results are the opposite when TC increases. Therefore, the optimal TC must be found to achieve satisfactory performance at all wind speeds. For wind speeds greater than 12 m/s, there is no difference in C_p because power control is performed for the rated power.

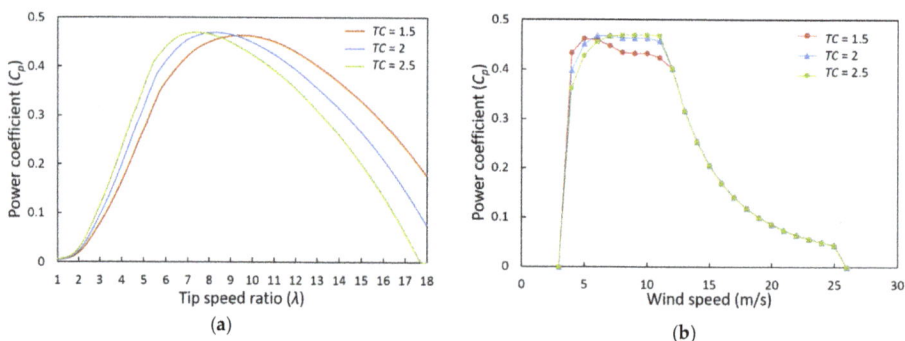

Figure 10. Aerodynamic performance according to the chord length: power coefficient versus (**a**) TSR and (**b**) wind speed.

The FAST results for three blades with different chord profile slopes (a_1 = 0.04, 0.06, and 0.08) are shown in Figure 11. The variation of C_p with the slope shows patterns similar to those in Figure 10. As an increase in chord slope also increases the blade size; the C_p–λ curve is shifted to low TSRs when the chord slope increases (Figure 11a). The slope, however, has less influence than the chord length (Figures 10b and 11b).

The FAST results to illustrate the differences in aerodynamic performance among three blades with different twist angle profiles (a_2 = 0.15, 0.2, and 0.25) are shown in Figure 12. The C_p–λ curve (Figure 12a) shows that there is no significant difference at low TSRs, and differences are observable at high TSRs. This finding indicates that the twist profile significantly affects the aerodynamic efficiency at low wind speeds. The graph of C_p versus wind speed (Figure 12b) shows that the low twist profile ($a_2 = 0.15$) corresponds to low C_p at low wind speeds (4–7 m/s), but it has similar values at higher wind speeds.

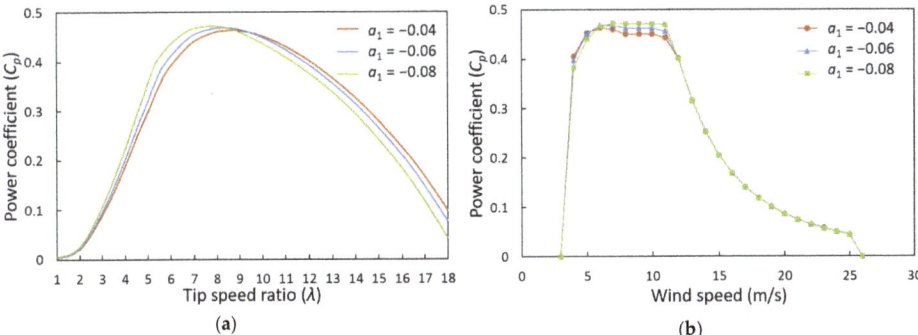

Figure 11. Aerodynamic performance according to the chord profile slope: power coefficient versus (a) TSR and (b) wind speed.

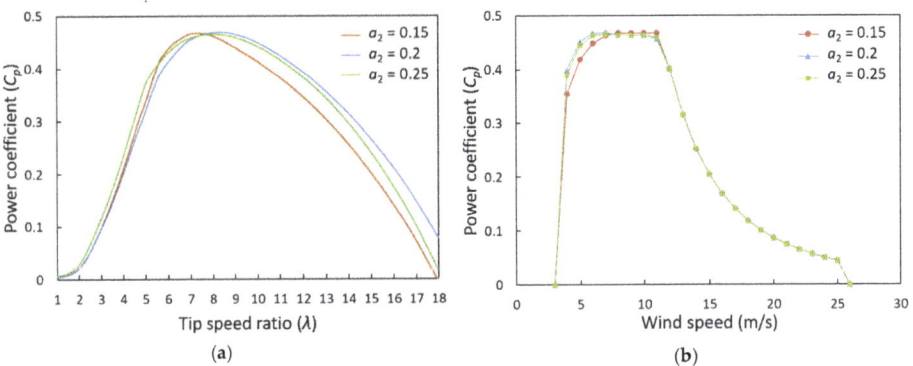

Figure 12. Aerodynamic performance according to the twist profile: power coefficient versus (a) TSR and (b) wind speed.

4. Wind Turbine Blade Design Optimization

This section describes the optimal blade linearization performed by applying the design ranges for chord and twist linearization given above. As mentioned above, the blade geometry has various effects on C_p at each TSR. Therefore, the purpose of optimal blade design is to find the blade geometry that can ensure optimal performance in all wind speed ranges in which a wind turbine may operate. To this end, the objective function that considers multiple TSRs, which was proposed in Section 2.4, was introduced. Table 3 summarizes the linearization coefficients selected in Section 3.2, the TSR range, and the optimal coefficients derived through optimization. Regarding the TSR range, $TSR = 6$–10 was applied to exclude low TSRs that do not significantly affect the aerodynamic performance at the rated wind speeds and considering low wind speeds.

Figure 13 shows the variations of the objective function value in the total performance process of the algorithm. These values improve and converge to a certain value, although they fluctuate frequently in the early stage. This behavior clearly shows the characteristics of the SA algorithm. Figure 14 shows the blade geometry and twisted section airfoil derived through optimization, and Table 4 summarizes the baseline and optimized chords and twist angles.

Table 3. Linearization coefficient ranges for blade geometry optimization.

Parameter	Lower Bound	Optimized	Upper Bound
Reference chord length (TC) (m)	1.5	2.19	2.5
Chord slope coefficient (a_1) (–)	−0.04	−0.0552	−0.08
Twist linear coefficient (a_2) (–)	0.15	0.178	0.25
Design TSR (λ) (–)	6	–	10

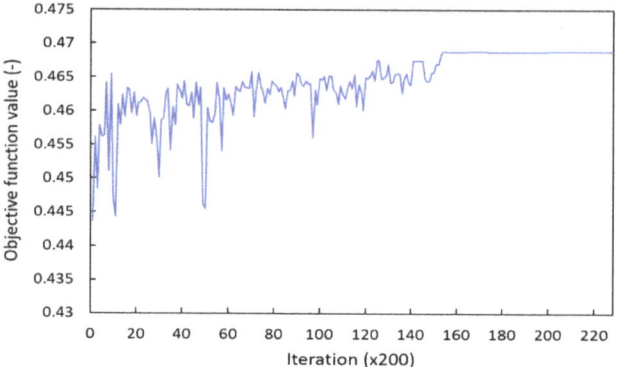

Figure 13. Variation of the objective function value in the optimization process.

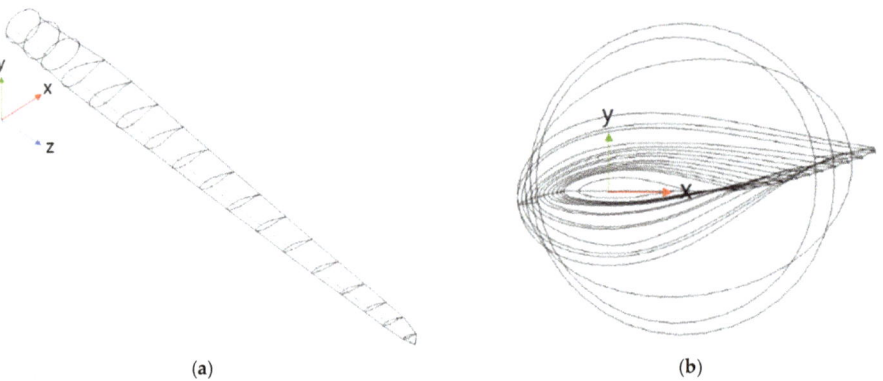

(a)   (b)

Figure 14. Optimized results: (a) blade geometry and (b) twisted section airfoils.

The chord and twist profiles of the blades derived through optimization with the NERL baseline are shown in Figure 15. In the chord profile (Figure 15a), there is no significant difference from the baseline, but the TC increases by 0.104 m and then slowly decreases toward the root section due to a reduction in the slope parameter b_1. As the overall size slightly decreases, it appears that the blade weight is lower than that of the baseline. In the twist profile (Figure 15b), the twist angle decreases by up to approximately 4° toward the root section compared to the baseline. The effects of these differences in geometry on the aerodynamic load of the blade are depicted in Figure 16, which shows the normal load (Figure 16a) and tangential load (Figure 16b) distributions of the optimized blade and baseline in the rotor radial direction. In the optimized case, the aerodynamic load in the tip region is lower than that of the baseline starting at a blade length of approximately 65%, but it is higher in the other sections. As long as the aerodynamic performance is not reduced, decreasing the aerodynamic load is favourable in terms of the structural aspects of the blade.

Table 4. Chord and twist distributions of baseline and optimized blades.

r/R	Baseline		Optimized	
	Chord (m)	Twist (°)	Chord (m)	Twist (°)
0.05	3.542	13.308	3.542	9.123
0.09	3.854	13.308	3.854	9.123
0.13	4.167	13.308	4.167	9.123
0.19	4.557	13.308	4.557	9.123
0.25	4.652	11.480	4.567	8.393
0.32	4.458	10.162	4.340	7.663
0.38	4.249	9.011	4.114	6.933
0.45	4.007	7.795	3.888	6.203
0.51	3.748	6.544	3.661	5.473
0.58	3.502	5.361	3.435	4.744
0.64	3.256	4.188	3.209	4.014
0.71	3.010	3.125	2.982	3.284
0.77	2.764	2.319	2.756	2.554
0.84	2.518	1.526	2.530	1.825
0.89	2.313	0.863	2.341	1.216
0.93	2.086	0.370	2.190	0.730
0.98	1.419	0.106	1.419	0.106

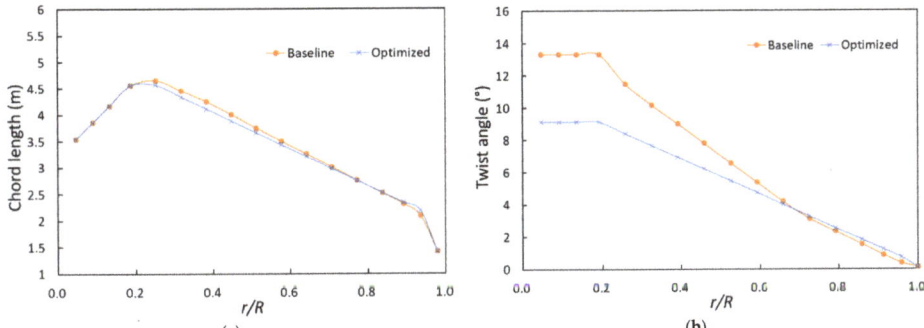

Figure 15. Chord and twist distributions of the baseline and optimized blades: (a) chord profiles and (b) twist angle profiles.

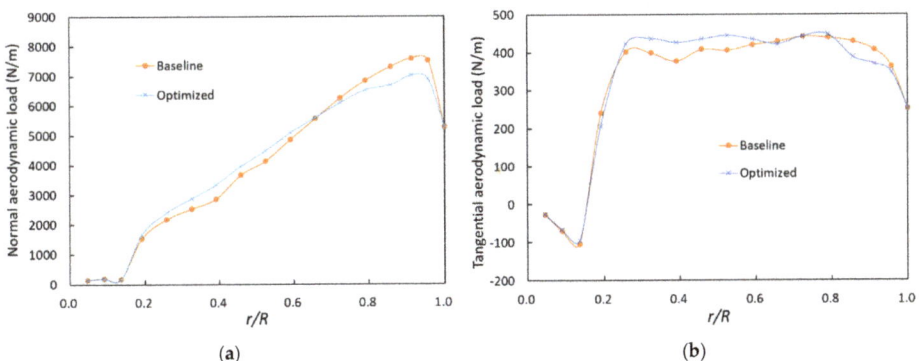

Figure 16. Aerodynamic load distributions of the baseline and optimized blades at a wind speed of 10 m/s: (a) normal load and (b) tangential load.

The aerodynamic performance of the optimized blade with the NREL baseline is shown in Figure 17. The C_p–λ curve (Figure 17a) shows that the C_p of the optimized blade is lower at low TSRs of 5–6 but higher when the TSR is 7 or higher. Thus, the wind turbine power can be improved at low wind speeds. The C_p versus wind speed graph (Figure 17b) shows that the optimized blade improves C_p at wind speeds of 4–6 m/s. In addition, the C_p is slightly improved at all wind speeds below the rated wind speed. As the NREL baseline also went through an optimization process in the design procedure, it may be difficult for new optimization to improve the efficiency significantly. However, the proposed optimization algorithm was verified by finding the optimal blade geometry that exhibited improved aerodynamic performance compared to the baseline. In addition, it was confirmed that the proposed objective function can produce blade designs with satisfactory aerodynamic efficiency at the targeted multiple TSRs.

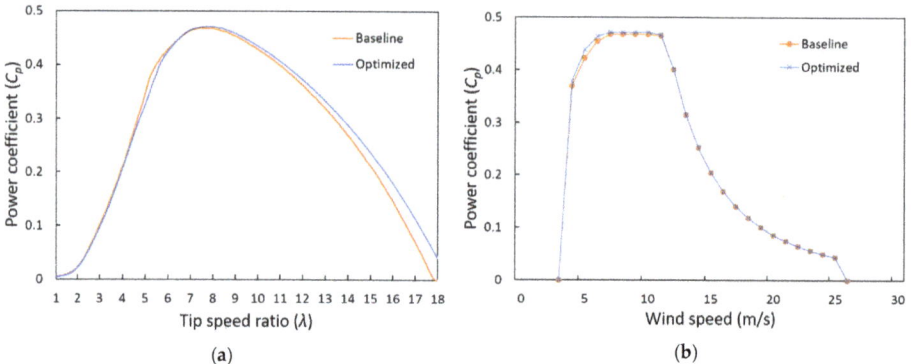

Figure 17. Aerodynamic performance of the baseline and optimized blades: power coefficient versus (a) TSR and (b) wind speed.

5. Conclusions

In this study, the aerodynamic performance of a wind turbine blade was examined according to the chord and twist linearization parameters, which are required in blade geometry design. In addition, an optimization algorithm for the linearization and an objective function that considers multiple TSRs to satisfy the optimal aerodynamic efficiency at various wind speeds were proposed. The main conclusions drawn by analysing the effects of the blade linearization parameters (chord length, chord profile slope, and twist profile) can be summarized as follows:

(1) In blade design, increasing the chord length increases the power coefficient at low wind speeds close to the cut-in wind speed but decreases the power coefficient towards higher wind speeds. Therefore, increasing the chord length is favourable for improving the aerodynamic efficiency at low wind speeds, but the design range must be selected considering the power coefficient reduction at high wind speeds and the weight caused by the chord length increase.
(2) Increasing the chord profile slope has an effect similar to that of increasing the chord length, but its effect on the aerodynamic efficiency is less than that of increasing the chord length.
(3) The twist profile mainly affects low wind speeds depending on the size of the incremental angle, and its effect on the aerodynamic performance at high wind speeds is not significant.

By applying the linearization parameter design ranges and multiple TSRs, a blade geometry that improved the power coefficient at all wind speeds below the rated wind speed compared to the NREL baseline wind turbine was derived. Therefore, it is necessary to consider multiple TSRs during blade design to satisfy the optimal aerodynamic performance at various wind speeds. As the structural performance of the blade was not addressed in this study, there is a limit to the practical blade design.

In the future, based on the results of this study, further research on blade geometry optimization considering blade structural stability for more practical blade design will be conducted.

Funding: This work was supported by the Korea Institute of Energy Technology Evaluation and Planning (KETEP) grant funded by the South Korean government (MOTIE) (NO. 20173010025010 and NO. 20184030202200).

Conflicts of Interest: The author has no conflict of interest.

References

1. Manwell, J.F.; McGowan, J.G.; Rogers, A.L. *Wind Energy Explained: Theory, Design and Application*; John Wiley & Sons, Ltd.: Chichester, UK, 2010.
2. Burton, T.; Jenkins, N.; Sharpe, D.; Bossanyi, E. *Wind Energy Handbook*, 2nd ed.; John Wiley & Sons, Ltd.: Chichester, UK, 2011.
3. Glauert, H. *Airplane Propellers, Aerodynamic Theory*; Springer: New York, NY, USA, 1963.
4. Wilson, R.E.; Lissaman, P.B.S. *Applied Aerodynamics of Wind Power Machines*; Oregon State University: Corvallis, OR, USA, 1974.
5. Wilson, R.E.; Lissaman, P.B.S.; Walker, S.N. *Aerodynamic Performance of Wind Turbines*; Oregon State University: Corvallis, OR, USA, 1976.
6. Johansen, J.; Madsen, H.A.; Gaunaa, M.; Bak, C.; Srensen, N.N. Design of a wind turbine rotor for maximum aerodynamic efficiency. *Wind Energy* **2009**, *12*, 261–273. [CrossRef]
7. Vaz, J.R.P.; Pinho, J.T.; Mesquita, A.L.A. An extension of BEM method applied to horizontal-axis wind turbine design. *Renew. Energy* **2011**, *36*, 1734–1740. [CrossRef]
8. Dai, J.C.; Hu, Y.P.; Liu, D.S.; Long, X. Aerodynamic loads calculation and analysis for large scale wind turbine based on combining BEM modified theory with dynamic stall model. *Renew. Energy* **2011**, *36*, 1095–1104. [CrossRef]
9. Lanzafame, R.; Messina, M. BEM theory: How to take into account the radial flow inside of a 1-D numerical code. *Renew. Energy* **2012**, *39*, 440–446. [CrossRef]
10. Yang, H.; Shen, W.; Xu, H.; Hong, Z.; Liu, C. Prediction of the wind turbine performance by using BEM with airfoil data extracted from CFD. *Renew. Energy* **2014**, *70*, 107–115. [CrossRef]
11. Maalawi, K.Y.; Badr, M.A. A practical approach for selecting optimum wind rotors. *Renew. Energy* **2003**, *28*, 803–822. [CrossRef]
12. Liu, X.; Wang, L.; Tang, X. Optimized linearization of chord and twist angle profiles for fixed-pitch fixed-speed wind turbine blades. *Renew. Energy* **2013**, *57*, 111–119. [CrossRef]
13. Yang, Z.; Yin, M.; Xu, Y.; Zhang, Z.; Zou, Y.; Dong, Z.Y. A multi-point method considering the maximum power point tracking dynamic process for aerodynamic optimization of variable-speed wind turbine blades. *Energies* **2016**, *9*, 425. [CrossRef]
14. Tahani, M.; Kavari, G.; Masdari, M.; Mirhosseini, M. Aerodynamic design of horizontal axis wind turbine with innovative local linearization of chord and twist distributions. *Energy* **2017**, *131*, 78–91. [CrossRef]
15. Yin, M.; Yang, Z.; Xu, Y.; Liu, J.; Zhou, L.; Zou, Y. Aerodynamic optimization for variable-speed wind turbines based on wind energy capture efficiency. *Appl. Energy* **2018**, *221*, 508–521. [CrossRef]
16. Selig, M.S. Application of a Genetic Algorithm to Wind Turbine Design. *ASME* **1996**, *118*, 22–28. [CrossRef]
17. Fuglsang, P.; Madsen, H.A. Optimization method for wind turbine rotors. *J. Wind Eng. Ind. Aerodyn.* **1999**, *80*, 191–206. [CrossRef]
18. Jureczko, M.; Pawlak, M.; Mężyk, A. Optimisation of wind turbine blades. *J. Mater. Process Technol.* **2005**, *167*, 463–471. [CrossRef]
19. Méndez, J.; Greiner, D. Wind blade chord and twist angle optimization using genetic algorithms. In Proceedings of the 5th International Conference on Engineering Computational Technology, Las Palmas de Gran Canaria, Spain, 12–15 September 2009; Civil-Comp: Stirlingshire, UK, 2009.
20. Polat, O.; Tuncer, I.H. Aerodynamic shape optimization of wind turbine blades using a parallel genetic algorithm. *Procedia Eng.* **2013**, *61*, 28–31. [CrossRef]
21. Tahani, M.; Sokhansefat, T.; Rahmani, K.; Ahmadi, P. Aerodynamic optimal design of wind turbine blades using genetic algorithm. *Energy Equip. Syst.* **2014**, *2*, 185–193.

22. Ashuri, T.; Zaaijer, M.B.; Martins, J.R.R.A.; van Bussel, G.J.W.; van Kuik, G.A.M. Multidisciplinary design optimization of offshore wind turbines for minimum levelized cost of energy. *Renew. Energy* **2014**, *68*, 893–905. [CrossRef]
23. Vianna Neto, J.X.; Guerra Junior, E.J.; Moreno, S.R.; Hultmann Ayala, H.V.; Mariani, V.C.; dos Santos Coelho, L. Wind turbine blade geometry design based on multi-objective optimization using metaheuristics. *Energy* **2018**, *162*, 645–658. [CrossRef]
24. Hansen, M.O.L.; Sørensen, J.N.; Voutsinas, S.; Sørensen, N.; Madsen, H.A. State of the art in wind turbine aerodynamics and aeroelasticity. *Prog. Aerosp. Sci.* **2006**, *42*, 285–330. [CrossRef]
25. Brownlee, J. *Clever Algorithms: Nature-Inspired Programming Recipes*; Lulu Press: Morrisville, NC, USA, 2011.
26. Jonkman, J.; Butterfield, S.; Musial, W.; Scott, G. *Definition of a 5-MW Reference Wind Turbine for Offshore System Development*; National Renewable Energy Lab. (NREL): Golden, CO, USA, 2009.

© 2020 by the author. Licensee MDPI, Basel, Switzerland. This article is an open access article distributed under the terms and conditions of the Creative Commons Attribution (CC BY) license (http://creativecommons.org/licenses/by/4.0/).

Article

A Control Scheme with the Variable-Speed Pitch System for Wind Turbines during a Zero-Voltage Ride Through

Enyu Cai [1,2,*], Yunqiang Yan [3], Lei Dong [1] and Xiaozhong Liao [1]

1. School of Automation, Beijing Institute of Technology, No. 5 South Zhongguancun Street, Beijing 100081, China; leidong@bit.edu.cn (L.D.); liaoxiaozhong@bit.edu.cn (X.L.)
2. Department of New Energy, GD Power Development Co. Ltd., Beijing 100101, China
3. GD Power Inner Mongolia New Energy Development Co. Ltd., Hohhot 010040, China; yunqiang.yan@chnenergy.com.cn
* Correspondence: 3120170446@bit.edu.cn or enyu.cai@chnenergy.com.cn or joshuacey@aliyun.com

Received: 3 May 2020; Accepted: 28 June 2020; Published: 30 June 2020

Abstract: Zero-voltage ride through (ZVRT) is the extreme case of low-voltage ride through (LVRT), which represents the optimal grid-connection capability of wind turbines (WTs). Enforcing ZVRT will improve the dynamic performance of WTs and therefore significantly enhance the resiliency of renewable-rich grids. A control scheme that includes a pitch system is an essential control aspect of WTs riding through voltage dips; however, the existing control scheme with a pitch system for LVRT cannot distinguish between a ZVRT status and a power-loss condition, and, consequently, does not meet the ZVRT requirements. A system-level control scheme with a pitch system for ZVRT that includes pitch system modeling, control logic, control circuits, and overspeed protection control (OPC) is proposed in this paper for the first time in ZVRT research. Additionally, the field data are shared, a fault analysis of an overspeed accident caused by a voltage dip that describes the operating status at the WT-collapse moment is presented, and some existing WT design flaws are revealed and corrected by the fault analysis. Finally, the pitching performance during a ZVRT, which significantly affects the ZVRT performance of the WT, is obtained from laboratory and field tests. The results validate the effectiveness of the proposed holistic control scheme.

Keywords: control scheme; variable-speed pitch; wind turbine (WT); zero-voltage ride through (ZVRT); low-voltage ride through (LVRT); overspeed protection control (OPC)

---

1. Introduction

Currently, low-voltage ride through (LVRT) capability is necessary for grid-connected wind turbines (WTs) in most countries and regions. This capability is intended to maintain a WT connection to the grid for between several hundred milliseconds to a few seconds when the voltage at the point of common coupling (PCC) dips, thereby large numbers of WTs from disconnecting from the grid due to frequent voltage fluctuations in weak-grid areas. The extreme case of LVRT, termed zero-voltage ride through (ZVRT), is defined by the voltage at the PCC falling to 0%, which represents the optimal grid-connection capability of WTs. The standard grid code curves for LVRT and ZVRT are illustrated in Figure A1.

The control scheme for a WT during a fault ride through (FRT) mainly consists of control strategies for the converter and a control scheme with a pitch system. Zhu and Zhang et al. introduced control strategies of the rotor-side converter (RSC) for LVRT in [1–4]. Liu and Yang provided useful and simplified models of the grid-side converter (GSC) for LVRT in [4,5], although GSC control during a ZVRT seems ineffective because the zero-voltage (ZV) dip causes the GSC to be essentially short-circuited. Some important equations for the transient analysis of the DFIG during a LVRT are given in [6–8]. Cheng adds more details regarding the transient analysis in [9] and even describes each step of the transient analysis procedure in various conditions. The equations to analyze rotor current during a crowbar action are necessary to evaluate the performance of a DFIG during a FRT. Most of the equations for the transient analysis of the DFIG in [6–9] were provided by Ouyang and Xiong in the book [10]. Some additional parameters, such as the electromagnetic torque as well as the d-axis and q-axis rotor currents, were collected from laboratory tests in [1,11] and are useful for reference.

Until now, however, few papers and results have been published regarding ZVRT. The only advances in ZVRT research are presented in [12–15]. References [12,13] provided a comprehensive control strategy under rapid pitch angle control, multiple levels of protection circuit response and an improved excitation control strategy during ZVRT. Reference [14] focused on the issue of reactive current injection during ZVRT. The models in [12–14] can be refined into one or more models that possess increased sophistication and complexity, and the lack of data from field tests could also be improved. Overall, the availability of published papers and results regarding ZVRT is still limited.

Zhang, Dou, and Burton et al. introduced the general physical structure of a variable-speed pitch system for large-scale WTs [16–18]. Although the transmission and converter of every type of large-scale WT have different physical structures, the variable-speed pitch systems in the WTs have similar physical structures and control principles. However, there have been improvements to these general physical structures in recent years. For example, the cabinet layout in every pitch subsystem has changed from a 7-cabinet style to a 3-cabinet style, making the physical structure of the pitch system more compact and reasonable.

Zhang presented a torque control modeling method for WTs [19]. Reference [17] presented a linearized-modeling method of the aerodynamic system for WTs. These modeling methods provide a premise for simulating a pitch system. The model must include the blades, a pitch angle control strategy, pitch actuators, and the transmission of WTs in order to validate a control scheme for a pitch system. In [20], a pitch angle control strategy based on fuzzy logic for variable-speed WTs was proposed. The pitch angle control strategy, including the pitch servo system, can be referenced and cited to model a pitch system. Reference [21] designed optimal output feedback controllers for a linear model of a WT at different operating points optimized by a genetic algorithm. The process of comparing these proposed controllers with a well-tuned proportional integral (PI) controller is interesting, but the practicability and feasibility of such a replacement for operating WTs still need to be proven. Chen presented a robust controller that adopts adaptive dynamic programming based on reinforcement learning and system state data in [22]. The pitch variation commands from this controller are relatively gradual, which reduces the energy consumption of the pitch actuator. Tang proposed an active power control strategy that integrated rotor speed and pitch angle regulation; this approach avoids the need for frequent pitch actuator actions while sustaining dispatched active power [23]. However, in practice, the reason given in [22,23] for reducing pitch actuator consumption may not be recommended. Timely pitching is the main requirement for WTs because that aspect is a basic safety rule; in contrast, the consumption of the pitch system is not a pressing issue for WT designers. Considering the action delay of mechanical devices in a pitch system after a controller outputs pitching commands, it is strongly suggested that the variations in the command effect timings should closely mirror the actual conditions.

A limited-angle torque electromechanical actuator for low-speed micro wind turbines used for overspeed protection is presented in [24]. The approach of improving overspeed protection control (OPC) performance by adding hardware devices is widely used in this field. The design idea in this article can be found in the OPC design of WTs. References [25,26] introduced the application of OPC in nuclear power plants. In [25], different OPC strategies can be obtained by switching the trigger signals or by selectively controlling the valves. The OPC system of WTs is not excessively complicated, so the designs can be simplified. Thus, the OPC program in [26] can be partially applied to the OPC design of WTs. In [27], a dynamic governor system model with a deflector control based on a pelton turbine, whose role is similar to that of thermal power overspeed protection, is developed. The purpose of this technology was to govern speeds but was not intended as a safety design. The controller based on the Proportional-Integral-Derivative (PID) governor model might be utilizable in WT converters. Mellal et al. addressed the availability and cost of an overspeed protection system for a gas power plant [28]. OPC availability is quantified, and an availability and cost curve is presented in this article. This issue is very interesting, and various types of power plants will have similar needs.

The method of testing OPC for thermal turbines was proposed by IEEE [29]. The suggested field test method for LVRT, along with data processing formulas and some technical details regarding field tests, is provided by IEC and IEEE [30,31]. However, to be usable for ZVRT, the test equipment needs to be improved and imbued with this capability. The mobile equipment in Standard [30] cannot satisfy the test requirements. Stationary equipment may be more applicable to ZVRT field tests. Standards [32,33] in Australia provide a reasonable and sufficiently strict ride-through time, which is very important for ZVRT field tests.

According to the literature review, the existing results have a disadvantage that can be summarized as follows: no control scheme with a variable-speed pitch system exists for WTs in ZVRT research. However, a control scheme that includes a pitch system is very necessary. No type of variable-speed-pitch WT could ride-through a ZV dip without a control scheme for a pitch system. Furthermore, the control scheme with a variable-speed pitch system for LVRT does not satisfy the ZVRT requirements because, when a WT detects that the voltage at the PCC has dipped to 0%, the WT controller will identify the WT as having lost its power supply. Then, an emergent fault status will be triggered, and the pitch system will continue pitching towards a feather in this emergent mode until the WT stops and self-locks. Obviously, a WT using this scheme cannot ride-through the ZV dip condition.

Therefore, the paper proposes a holistic control scheme with a pitch system for WTs during a ZVRT that can satisfy the ZVRT requirements to improve the grid-connection capability of WTs.

The contributions of this paper are as follows:

(1) proposal of a system-level control scheme with an electric variable-speed pitch system for ZVRT in WTs that includes pitch system modeling, control logic, control circuits, and OPC, which is first in ZVRT research;

(2) presentation of a fault analysis of an overspeed accident caused by a voltage dip that describes the operating status at the moment of WT collapse along with presentation of the associated field data;

(3) correction of design flaws in the existing control scheme for LVRT that improves the fault response performance of a pitch system;

(4) completion of a ZVRT field test based on an operating WT to verify the effectiveness of the holistic control scheme with the variable-speed pitch system.

The remainder of the paper is organized as follows: Section 2 shows the physical structure of the variable-speed pitch system. Section 3 explains the modeling method of a pitch system including the pitch angle control strategy. Section 4 introduces the control logic and control circuits for LVRT and ZVRT. Section 5 designs a new OPC system for ZVRT in WTs. Section 6 presents the results of the simulation, the laboratory test, and the field test to validate the effectiveness of the holistic control scheme. Finally, Section 7 concludes and discusses future work.

2. Physical Structure of a Variable-Speed Pitch System

Although every type of WT has a different physical structure, the variable-speed pitch systems of WTs have a similar physical structure. Figure 1a,b show that the pitchable blades absorb wind energy and drive the hub and the shaft (and, ultimately, the generator) around the longitudinal axis. A pitch system in a WT is composed of several separated subsystems. Each pitch subsystem, which is installed in the hub and corresponds to a blade, has the same structure (shown in Figure 1c) and function. A pitch subsystem has a servo motor and a reducer to drive the driving pinion that engages with the inner ring of the pitch bearing. The pitch bearing at the joint of the blade root and the hub implement adjust the pitch angle of the blade, which is essential in controlling the power of WTs [16].

Figure 1d introduces the overall structure of a pitch subsystem. Both the WT power and the batteries (or capacitors, as the backup power) provide direct current (DC) power and actuate the servo motor through a pitch actuator. The WT power can actuate to pitch both towards feather and fine, but the backup power can actuate to pitch towards feather only. The WT controller controls every pitch controller and exchanges data with them. The pitch controller, actuator, servo motor, and encoder form a control loop. The position sensor provides a correction for the encoder in the control.

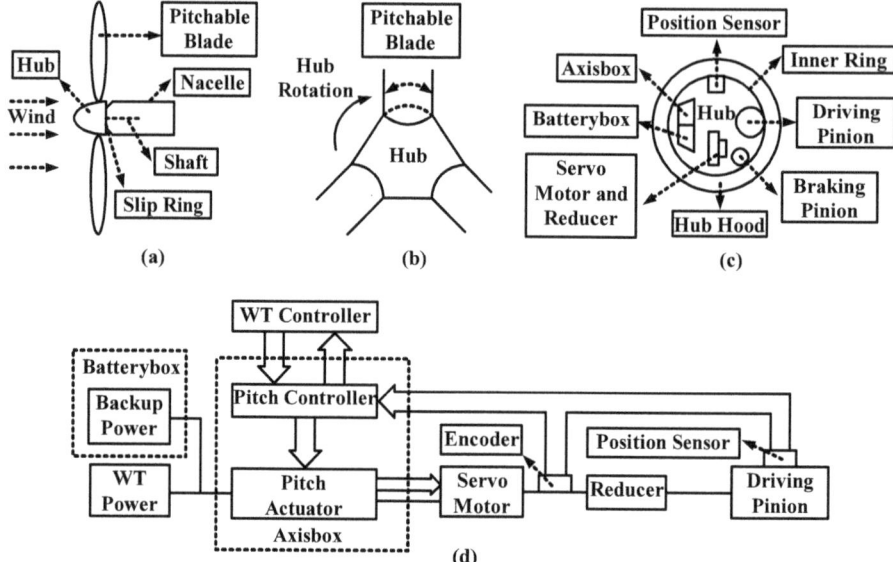

Figure 1. Physical structure of the variable-speed pitch system: (a) side view of the hub; (b) front view of the hub; (c) perspective view of the hub; (d) schematic diagram of a pitch system.

3. Modeling and Variable-Speed Pitch Control

Theoretically, pitch control with a variable-speed pitch system for WTs is the process that adjusts the pitch angles of the blades. As shown in Figure 1d, the WT controller sends the reference pitch angles to the pitch controllers. To simulate the process, a WT model that includes both the aerodynamic components and the pitch control strategy is necessary. The modeling parameters are shown in Table 1.

Table 1. Modeling parameters.

Parameter Category	Value	Parameter Category	Value
Rotor diameter	82.76 m	Maximum torque of the driver	75 N·m
Rotor swept area	5384 m²	Rated torque of the driver	28.7 N·m
Rotor rated speed	17.4 rpm	Rated power of the driver	4.5 kW
Rotor speed range	9.7 rpm–19.5 rpm	Rated voltage of the driver	29 V
Design tip speed ratio	8.5	Brake torque of the pitch	100 N·m
Gear ratio	1:100.48	Maximum pitch speed	8.3 deg/s
Pitch type	Electric pitch	Pitch response time	100 ms

3.1. Wind Turbine Modeling

For an actual torque, the rotor speed can be maintained in proportion to the wind speed, which means that the optimum tip speed ratio is maintained, and the power coefficient C_p is a maximum. The power coefficient indicates the fraction of the potential power in the wind, which can be converted by the WT. It has a theoretical maximum value of 59.3% (the Betz limit). In practice, perfectly tracking the optimum C_p is always a target for the control performance of the pitch system. The aerodynamic model can be expressed as (1) and (2) [19]:

$$P = \frac{1}{2}\rho \pi R^2 C_p(\lambda, \beta) v_{wind}^3 \tag{1}$$

$$T_a = \frac{1}{2}\rho \pi R^3 \frac{C_p(\lambda, \beta)}{\lambda} v_{wind}^2 \tag{2}$$

where P is the output power, T_a is the aerodynamic torque, ρ is the air density, R is the radius of a blade, v_{wind} is the wind speed, β is the pitch angle, and λ is the tip speed ratio.
$C_p(\lambda, \beta)$ is given by (3) and (4) [19].

$$\lambda = \frac{\omega_r R}{v_{wind}}, \lambda_i = (\frac{1}{\lambda - 0.02\beta} - \frac{0.003}{\beta^3 - 1})^{-1} \tag{3}$$

$$C_p(\lambda, \beta) = 0.73(\frac{151}{\lambda_i} - 0.58\beta - 0.002\beta^{2.14} - 13.2)e^{-\frac{18.4}{\lambda_i}} \tag{4}$$

where ω_r is the angular speed of the rotor.

The dynamic response of ω_r can be expressed as (5) [17]:

$$J_r \dot{\omega}_r = T_a - T_s - D_r \omega_r, \tag{5}$$

where J_r, $\dot{\omega}_r$, T_s and D_r denote the rotational inertia of the rotor, the accelerated speed of the rotor, the torque of the main shaft, and the damping coefficient of the rotor, respectively.

According to (2) and (5), $J_r \dot{\omega}_r$ is a nonlinear function of ω_r, v_{wind} and β, as shown in (6) [17]:

$$J_r \dot{\omega}_r = f(\omega_r, v_{wind}, \beta). \tag{6}$$

According to (2) and (6), the WT model is nonlinear. A linearized model is necessary for the pitch controller. At the optimum point, the Taylor series expansion of (6) can be expressed as (7) [17]. The subscript op denotes at the optimum point:

$$J_r \dot{\omega}_r = f(\omega_{r\_op}, v_{wind\_op}, \beta_{op}) + [\frac{\partial f}{\partial \omega_r}\Delta\omega_r + \frac{\partial f}{\partial v_{wind}}\Delta v_{wind} + \frac{\partial f}{\partial \beta}\Delta\beta]$$
$$+ \frac{1}{2!}[\frac{\partial^2 f}{\partial \omega_r^2}(\Delta\omega_r)^2 + \frac{\partial^2 f}{\partial v_{wind}^2}(\Delta v_{wind})^2 + \frac{\partial^2 f}{\partial \beta^2}(\Delta\beta)^2 + 2\frac{\partial^2 f}{\partial \omega_r \partial v_{wind}}\Delta\omega_r \Delta v_{wind} + 2\frac{\partial^2 f}{\partial v_{wind} \partial \beta}\Delta v_{wind}\Delta\beta + \tag{7}$$
$$2\frac{\partial^2 f}{\partial \omega_r \partial \beta}\Delta\omega_r \Delta\beta] + ...$$

where $\Delta\omega_r = \omega_r - \omega_{r\_op}$, $\Delta v_{wind} = v_{wind} - v_{wind\_op}$, $\Delta\beta = \beta - \beta_{op}$.
Equation (3) deduces (8) [17].

$$\frac{\partial \lambda}{\partial v_{wind}} = \frac{\omega_r R}{v_{wind}^2} = -\frac{\lambda}{v_{wind}} \tag{8}$$

Supposing $f(\omega_{r\_op}, v_{wind\_op}, \beta_{op}) = 0$, (7), and (8) can reduce to (9) [17].

$$J_r \dot{\omega}_r = \frac{\partial f}{\partial \omega_r}\Big|_{op} \cdot \Delta \omega_r + \frac{\partial f}{\partial v_{wind}}\Big|_{op} \cdot \Delta v_{wind} + \frac{\partial f}{\partial \beta}\Big|_{op} \cdot \Delta \beta \tag{9}$$

Supposing $\gamma = \frac{\partial f}{\partial \omega_r}\Big|_{op}$, $\zeta = \frac{\partial f}{\partial v_{wind}}\Big|_{op}$, and $\xi = \frac{\partial f}{\partial \beta}\Big|_{op}$, (10) can be obtained and s is the Laplace operator. Then, (11) is deduced by (10) as a linearized model according to the Laplace transform [17]:

$$J_r s \Delta \omega_r(s) = \gamma \Delta \omega_r(s) + \zeta \Delta v_{wind}(s) + \xi \Delta \beta(s) \tag{10}$$

Supposing $\eta = \gamma / J_r$,

$$\Delta \omega_r(s) = [\frac{\zeta}{J_r} \Delta v_{wind}(s) + \frac{\xi}{J_r} \Delta \beta(s)] \cdot \frac{1}{s-\eta}. \tag{11}$$

3.2. Pitch Servo System

Figure 2 illustrates a typical pitch control system, including the PI controller, pitch servo, and a nonlinear turbine model [20]. The pitch servo system consists of the actuator and motor shown in Figure 1d. The pitch actuator is a nonlinear servo that generally rotates all of the blades or a part of them. In the closed loop, the pitch servo can be modeled as an integrator or a first-order delay system with a time constant τ_c. Equation (12) provides the dynamic behavior of this system [20]:

$$\frac{d\beta}{dt} = -\frac{1}{\tau_c}\beta + \frac{1}{\tau_c}\beta_{ref} \tag{12}$$

where $\beta_{min} \leq \beta \leq \beta_{max}$, $(\frac{d\beta}{dt})_{min} \leq (\frac{d\beta}{dt}) \leq (\frac{d\beta}{dt})_{max}$, and the subscripts min, max, and ref denote the minimum, maximum, and reference values, respectively.

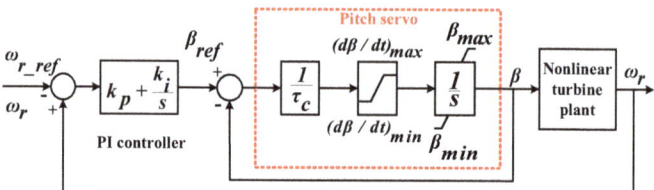

Figure 2. Block diagram of a typical pitch control system.

3.3. Pitch Angle Control Strategy

Figure 3 illustrates the pitch control system with a linearized turbine model. The PI controller receives the reference and measured turbine rotational speed, that is, ω_{ref} and ω_r, respectively, calculate the control error $\omega_r - \omega_{ref}$ and generate the pitch angle reference signal β_{ref}. The pitch servo is also a closed-loop system that can rotate the pitch angle β to track the reference signal. The adjusted pitch angle will move the turbine rotational speed to the reference value, and therefore eliminate the control error.

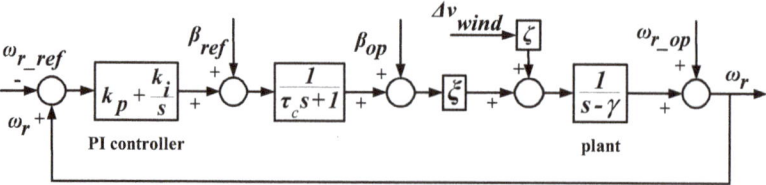

Figure 3. Block diagram of the pitch angle control strategy.

Figure 3 reveals the denominator of the transfer function for $\omega_r(s)$ as (13) [20]:

$$H(s) = \frac{s(s + \frac{1}{\tau_c})(s - \gamma) - \xi(sk_p + k_i)}{s(s + \frac{1}{\tau_c})(s - \gamma)} \qquad (13)$$

For stability, the components of the terms in the transfer function need to be positive. If the optimum point is changed, the PI controller gains should be redesigned to maintain the dynamic response and stability of the system.

4. Control Scheme for ZVRT

By definition, the voltage dip at the PCC is a fault for WTs. However, the ride-through process requires WTs to not disconnect from the grid, based on the premise that the WTs are running safely. Therefore, the control scheme during a ZVRT or an LVRT for a WT is actually a safety control scheme in the fault status.

4.1. Design of Control Logic during an FRT

There are three control modes in the WT controller: normal, fast, and emergent modes; the latter two of which are used to deal with faults. In fast mode, the pitch system pitches towards feather. In emergent mode, the WT controller outputs an emergency feather command (EFC). The pitch system will not only feather to the limit but also ignore other commands until the WT stops and selflocks.

Many fault indicators are defined in the WT controller to reflect the operating status of the WT. Any high-level output from the indicators triggers the fault status, and the WT controller switches fast or emergent mode. The logic is shown in Figure 4. In particular, a turn-off delayer (TOF) is added after the OR logic operation, which contributes to maintaining the input of the TOF for 4 s. The sampling period in the WT controller is 20 ms. A total time of 4 s or 200 periods aims to ensure that there is sufficient time for communication, signal processing, and device action when the OR logic operation outputs a high level.

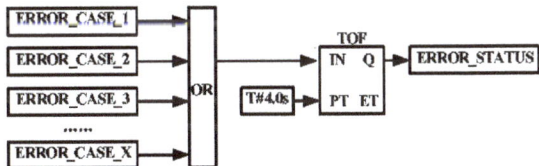

Figure 4. The control logic of triggering the fault(error) status for WTs.

Different from other fault indicators, the 18 error variables shown in Table 2 indicate that the WT has been operated in an emergency and must be stopped and checked as soon as possible. As shown in Figure 5, the variables (from V1 to V18) control the nodes (from K1 to K18) of a safe loop. Any output variation of the variables indicates the breakup of a node, which would lead to an interruption of the loop and the emergent mode in the WT controller. The threshold values of the variables are shown in Table 3.

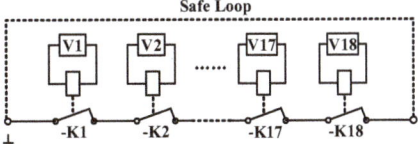

Figure 5. Schematic diagram of the safe loop.

Table 2. Error variables to interrupt the safe loop.

Fault Category	Fault Name
Overspeed	Overspeed error in rotor
Yaw	Cable twist error
Vibration	Vibration error in nacelle
Others	Stop command disabled error
Controller	Safety system controller error
Controller	Safety system program error
Controller	Manual emergent stop
Controller	Overspeed protection error
Controller	Programmable logic controller error
Pitch	Pitch position error
Pitch	Pitch communication error
Pitch	Pitch battery error
Pitch	Pitch battery charger error
Pitch	Pitch converter error
Pitch	Direct current monitoring error in the pitch converter
Pitch	Power supply fuse error
Pitch	Interrupting safe loop request from the pitch controller
Pitch	Emergency status in the pitch system

Table 3. Threshold values of the fault variables.

Fault Variable	Threshold Value
Limit rotor speed in fast mode	19.5 rpm
Limit rotor speed in emergency mode	20.35 rpm
Detection D-value of the rotor speed	1.3 rpm
Nacelle vibration acceleration	0.15 g
Limit yaw position of the unwinding cable	690 deg
Limit yaw position	750 deg
Minimum yaw speed	0.2 deg/s
Detection D-value of the pitch angle	2 deg
Delay time for the maximum detection difference of the pitch angle	2 s
Angle of the starting pitch	50 deg
Upper limit temperature of the pitch battery box	65 °C
Lower limit temperature of the pitch battery box	−10 °C
Upper limit temperature of the pitch converter	90 °C
Lower limit temperature of the pitch converter	0 °C
Upper limit temperature of the pitch motor	140 °C
Lower limit temperature of the pitch motor	−30 °C
Maximum pitch speed	10 deg/s
Maximum pitch acceleration	20 deg/s^2

For the pitch system, the three modes in a WT controller correspond to three feather methods (normal-feathering, fast-feathering, and emergent-feathering), which can be distinguished by the pitching (towards feather) speeds (4.0, 5.5 and 7.0 deg/s) and the feather limit (pitch angles: 89°, 89° and 91°). The normal-feathering method is used for a WT in normal operation. When the WT switches the fault status, the pitch system will adopt the fast-feathering method. The pitch system receives the control commands from the WT controller periodically in the normal-feathering and fast-feathering methods and adjusts the pitching actions immediately according to the commands. However, once the emergent mode is confirmed by the WT controller, the pitch system ignores all the commands, continues pitching towards feather until the pitch angle reaches 91°, and can only be unlocked manually, which means that the WT may have critical failures and should be analyzed carefully.

There are two types of power supplies for the pitch controllers of the WT power: DC 24 V and an uninterrupted power supply (UPS, DC 24 V). In normal mode, the pitch controller uses the WT power: DC 24 V, and the pitch actuator including the servo motor uses the WT power. In emergent mode, the pitch controller uses the UPS, and the pitch actuator uses the backup power. In fast mode, the pitch system generally uses the WT power unless the voltage of the WT power is lower than a threshold value. However, this fault will trigger more WT faults.

As mentioned previously, the control process by the control logic for LVRT (the existing design) can be described as follows. Once a low-voltage (LV) dip at the PCC is detected, the WT controller outputs an LVRT-starting signal, and the pitch system executes the fast-feathering method. During the LVRT, if the PCC voltage returns to normal, the WT controller outputs an LVRT-ending signal and a pitch angle based on Section 3.3; then, the pitch system pitches in normal mode. In this period, the pitch system generally runs on the WT power, the voltage of which is detected periodically.

However, the existing design will not work for the ZVRT condition. Once a ZV dip occurs, the WT controller does not recognize it as a clear LV dip, but cannot determine if it is a power-loss case, and thus, the safe loop is interrupted almost simultaneously.

Therefore, to ride through, the control scheme during a ZV dip has been improved. First, the voltage dips at the PCC are differentiated. Voltages at the PCC between 5% p.u. and 90% p.u. are regarded as an LV dip. When a ZV dip (the voltage at PCC is between 0% p.u. and 5% p.u.) is detected, the pitch system still acts by the fast-feathering method as with an LVRT within a certain time. The time is set to 1000 ms because the model in Article [12] theoretically simulates the maximum time of a ZVRT as nearly 950 ms. After this time, the uncleared ZV dip is regarded as a power-loss case, and the WT controller outputs an EFC. Second, to avoid the power-loss risk, the UPS and backup power are enabled as power supplies for the pitch controller and pitch actuator during a ZVRT. Third, the OPC system of WTs is redesigned to avoid the overspeed error in the rotor caused by untimely pitching.

4.2. Design of Control Circuit during an FRT

The control circuit is designed to execute the control logic in Section 4.1, which involves the WT controller, pitch controller, pitch actuator, and pitch power in Figure 1. For the pitch system, a major function of the control circuit is switching or selecting the power for the pitch actuator.

4.2.1. Control Circuit for LVRT

The LVRT control circuit is shown in Figure 6. Figure 6a introduces the control structure of a DC contactor. Contacts 1–4 are the main contacts. Contacts 11, 12, and 14 are auxiliary contacts. Contact 11 is a common port. Contact 12 is a normally closed (NC) contact. Contact 14 is a normally open (NO) contact. When Coils A1 and A2 operate, Contacts 1–4, 11, and 14 are synchronized to operate. Contact 12 executes a reverse action. Elements 6K3 and 5K1 in Figure 6b are DC contactors. 304S1 is a manual switch. Module KL2134 is a 4-channel digital output terminal for electrical isolation, which provides binary control signals [34]. KL2134 will output a high level when a combination of the following conditions is true. (1) The pitch angle reaches 91°. (2) The pitch actuator is fault-free. (3) The WT controller outputs a reset command. KL2134 and 304S1 together provide a bypass circuit. Figure 6c is the power-switching circuit of the pitch actuator (including the servo motor).

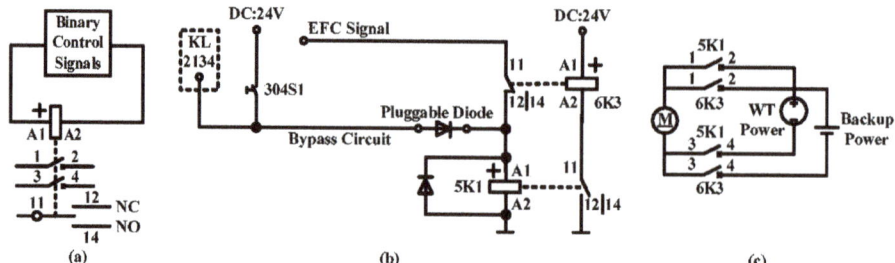

Figure 6. Control circuit for LVRT. (a) schematic diagram of DC contactor; (b) control circuit for LVRT; (c) power switching circuit.

The control process can be described as follows. In normal mode, the EFC signal continuously outputs a high level. Contacts 11 and 12 of 6K3 operate. The A1 side of 5K1 receives the high level and Contacts 1–4 of 5K1 operate. At the same time, Contacts 11 and 12 of 5K1 release, which keeps A1, A2, and Contacts 1–4 of 6K3 released. The WT power is enabled for the pitch actuator.

When the safe loop is interrupted, the EFC signal outputs a low level. However, at the moment of interruption, Contacts 11 and 12 of 6K3 keep operating, which triggers A1 and A2 of 5K1 release. Then, Contacts 11 and 12 of 5K1 operate, and Contacts 1–4 of 5K1 release. Next, A1, A2, and Contacts 1–4 of 6K3 operate, and Contacts 11 and 12 of 5K1 release. Finally, the backup power is enabled, and the pitch system starts to feather in emergent mode. Using this design, 5K1 and 6K3 implement a mutex in the control logic.

As mentioned above, a safe loop interruption can only be restored manually. After the WT stops, the backup power will be enabled. When the fault is cleared, the WT power needs to be enabled for pitching towards fine. The bypass circuit in Figure 6b is designed for this power switching requirement, which aims to temporarily span the EFC signal. When 304S1 is closed, the bypass circuit outputs a high level to make A1 and A2 of 5K1 operate and finally enable the WT power (A1 and A2 of 5K1 operating causes Contacts 11 and 12 of 5K1 to release and Contacts 1–4 of 5K1 to operate. Then, A1 and A2 and Contacts 1–4 of 6K3 release.). KL2134 is actually a redundancy design for the bypass circuit. If it is verified that there is no fault for the WT and the emergent mode is triggered just because an EFC signal outputs incorrectly, KL2134 provides a remote-restoring method for the WT by resetting the WT controller. Conditions (1) and (2) for KL2134 outputting a high level particularly limit the scope of application of KL2134, which means that KL2134 outputs a high level only if the blade has reached a feather limit and there are no faults in the pitch system.

4.2.2. Flaw Analysis of Control Circuit for LVRT

A WT operating in Liaoning, China, collapsed on 6 May 2017, due to an LV dip. The cause analysis indicates conclusively that there were design flaws in the LVRT control circuit. The analysis process can be described as follows.

Figure 7 demonstrates the operating status of the WT at the moment of collapse based on data in the emergent record of the WT. The pitch angle of Blade 1 and the A-phase grid voltage are presented as examples of the pitch angles (of three blades) and 3-phase grid voltages. For convenience, all the actual values are converted to the p.u. values.

As shown in Figure 7, the grid voltage decreased to 0.8 p.u. at the 26th second, which triggered an LVRT starting signal. The dip also caused a voltage fluctuation of the WT power (The emergent record showed that the WT power fluctuated between 182.39 V and 400 V). According to the LVRT control logic, the threshold value triggering the pitch system to switch to backup power during an LVRT was set as 115 V. Therefore, the backup power was not enabled for the voltage fault of the WT power during the collapse.

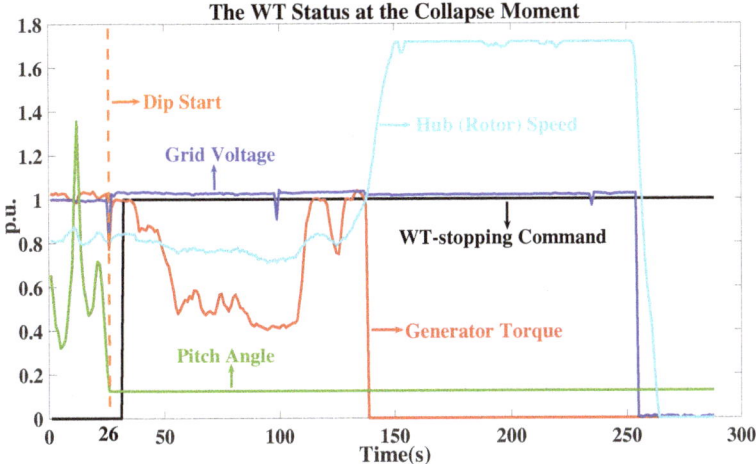

Figure 7. The operating status of the WT at the moment of collapse. The p.u. values of the rotor speed, generator torque, pitch position, and grid voltage are 20.35 rpm, 100 N·m, 10°, and 380 V, respectively.

Then, the WT controller outputted the pitch communication error in Table 2. After the safe loop was interrupted, the WT-stopping command outputted a high level. However, the pitch position revealed that the pitch system did not act, and the WT converter decreased the reference electromagnetic torque of the DFIG in a timely manner. Next, the rotor speed increased to more than 1.7 times the upper limit, and the WT controller output the overspeed error, grid disconnection error, high temperature of the winding in the generator, and high temperature of the bearing of the generator in this order. Finally, the transmission system of the WT disintegrated, the blades broke, and the WT went offline from the monitoring system. The collapse process lasted 250 s (after 250 s, most of the curves returned to zero, which meant that the WT had already collapsed), and the largest rotor speed of the generator was over 3500 rpm.

The immediate cause of the accident was the overspeed error in the rotor. The model in Section 3.1 is used to simulate an extreme case. Suppose that Blade 2 and Blade 3 cannot pitch due to faults. The results in Figure 8 indicate that the rotor speed would be held at a safety level even if there was only one blade that could be feathered. Then, it could be concluded that any one of the feathered blades would effectively decrease the rotor speed, which means the pitching load is reasonable.

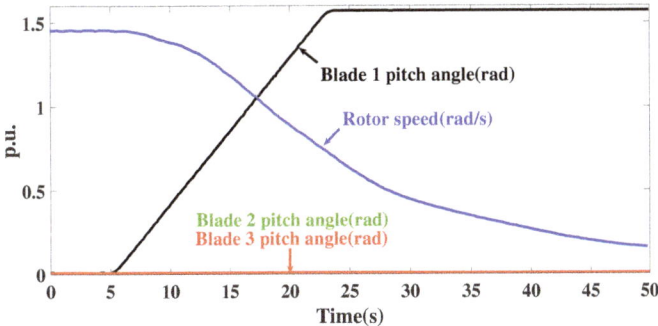

Figure 8. Simulation results of the extreme case of the WT in an accident. The p.u. values of the pitch angle and rotor speed are 58.20° and 9.3 rad/s, respectively.

To further analyze this example, all the wrecked devices of the pitch system in this accident were detected carefully. The key point occurred when KL2134 continuously outputted a high level because the inner signal channel short-circuited. As shown in Figure 6, if KL2134 outputs a high level incorrectly and continuously, A1, A2, and Contacts 1–4 of 5K1 continue operating, which prevents Contacts 1–4 of 6K3 from operating, and the backup power cannot be enabled. When the WT power could not provide the power for pitching during the LV dip, the pitch system thoroughly lost power for pitching and could not feather.

4.2.3. Improved Control Circuit for ZVRT

It is probable that inner damage occurred to all modules. Thus, the flaw is not a faulty KL2134 module that should be replaced but a design flaw in which the control circuit lacks safety redundancy.

The control circuit in Figure 6b is redesigned for the bypass circuit and the (signal) input part. As shown in the red box in Figure 9, DC contactors 1K1 and 2k3 are added in the bypass circuit, and their structures are shown in Figure 6a. Contacts 1–4 are the main contacts of 1K1. Module KL1104 is a 4-channel digital input terminal for electrical isolation, which detects the input level periodically as a monitor [35]. The conditions of KL2134-1 outputting a high level are identical with those of KL2134 in Figure 6b. However, the condition of KL2134-2 outputting a high level is that the WT controller provides a ZVRT-ending command. 208F9 is an air switch to isolate the UPS from the contacts.

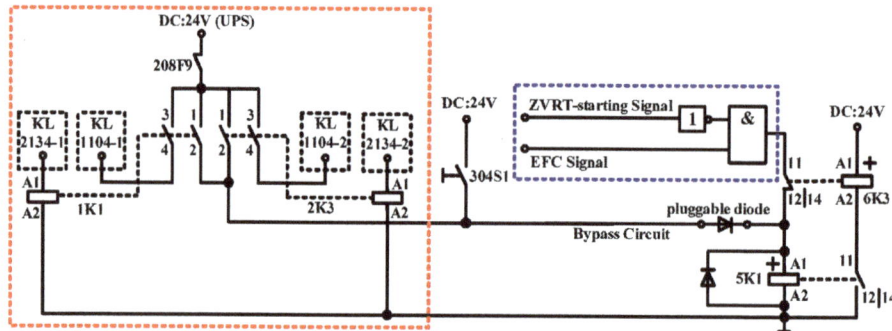

Figure 9. Improved control circuit for ZVRT.

When KL2134-1 outputs a high level according to the WT power command, A1, A2, and Contacts1–4 of 1K1 operate, and the bypass circuit outputs a high level from DC:24V (UPS). When KL2134-1 outputs a high level incorrectly and continuously, KL1104 will continuously detect the high level from DC:24V (UPS). To emphasize this point, both the WT controller command and fault will make KL2134-1 output a high level, but the reset command from the WT controller will be maintained for 4 s, which has the same design as that shown in Figure 4. KL1104 only detects a high level if the time is longer than 4 s, and it would output a high level for the WT controller, which means that KL2134-1 has been damaged.

KL2134-2 is added for the ZVRT-starting signal in the blue box in Figure 9. When the ZVRT-starting signal outputs a high level, the backup power will be enabled. However, when the ZVRT status ends, the WT power needs to be enabled for pitching towards fine. The WT controller outputs a ZVRT-ending signal that lasts for 4 s for KL2134-2, and KL2134-2 outputs a high level. 2K3 and KL1104-2 act in the same way as 1K1 and KL1104-1.

5. Overspeed Protection Control (OPC)

Section 4.1 indicates the importance of overspeed protection during a ZVRT. The overspeed protection during a ZVRT provides a necessary safety redundancy for WTs to respond to power-loss risks and not-pitching failures. The pitch system is exactly the executor of OPC.

The existing OPC scheme is shown in Figure 10. Sensors are installed in different locations in the nacelle and measure different types of rotor speeds, such as the shaft speed and the rotor speed in the generator. The OPC1 module is installed in the WT controller cabinet and receives the rotor speed data from the sensors. If any rotor speed is beyond the threshold value, the OPC1 module will send an overspeed signal to the WT controller. Then, the WT controller will interrupt the safe loop and output an EFC to the pitch controller. The EFC signal will output a low level to enable the backup power (explained in Figure 6) because the NC contact K1 turns to open. However, the OPC devices in the existing scheme are far away from the pitch system in the structure, which produces a substantial amount of long-distance signal transmission. Section 4.2.2 and Appendix B confirm that there may be some failures in the long-distance signal transmission, which will eventually lead to not-pitching failures. Furthermore, in the control logic during a ZVRT, the pitch system will pitch by the fast-feathering method in the first 1000 ms. If the ZV dip is exactly a power-loss situation, the pitch system might lose communication and power supply after 1000 ms, resulting in an overspeed risk.

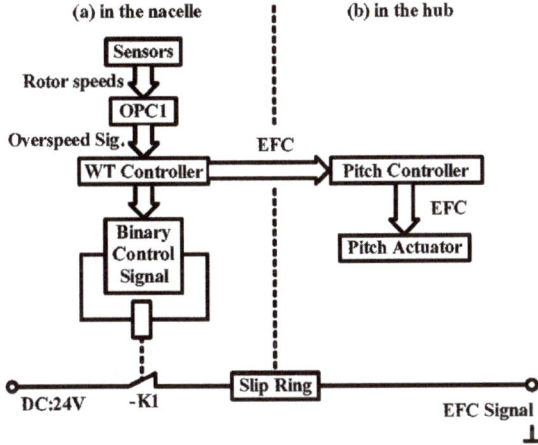

Figure 10. The existing OPC scheme.

Therefore, an improved OPC scheme is proposed, as shown in Figure 11. For the improved scheme, three sensors are installed in the hub to measure one type of rotor speed, the hub speed. A new OPC module (OPC2) separated from the WT controller is designed and installed in the pitch controller cabinet in the hub, which is directly connected with the pitch controller. The OPC2 module receives the rotor speed data from the three sensors. Once it outputs an overspeed signal, the pitch controller will interrupt the safe loop immediately, although the overspeed signal has been sent to the WT controller at the same time. The OPC2 module provides an additional protection for WTs, compared with the existing OPC scheme. The pitch controller in the improved scheme has a greater control authority, and the control process based on the OPC2 module will occur in the hub only, which can effectively reduce failures.

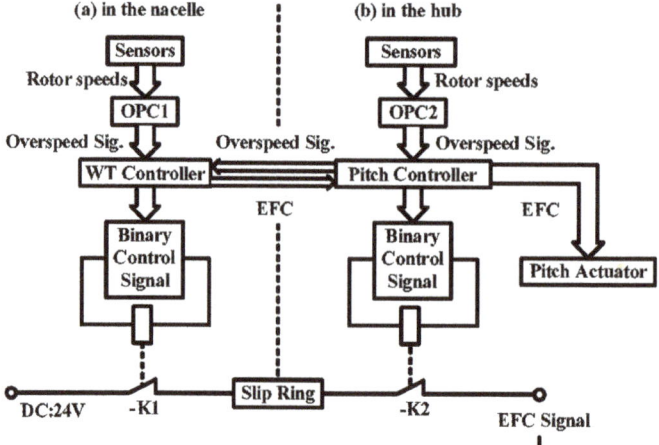

Figure 11. The improved OPC scheme.

The three photoelectric sensors (proximity switches) and corresponding limit irons used to locate the sensors need to be embedded in the hub. They are 120° apart from each other. The installation diagram is shown in Figure 12, and Figure 13 shows the installation of the OPC2 module in the pitch controller cabinet.

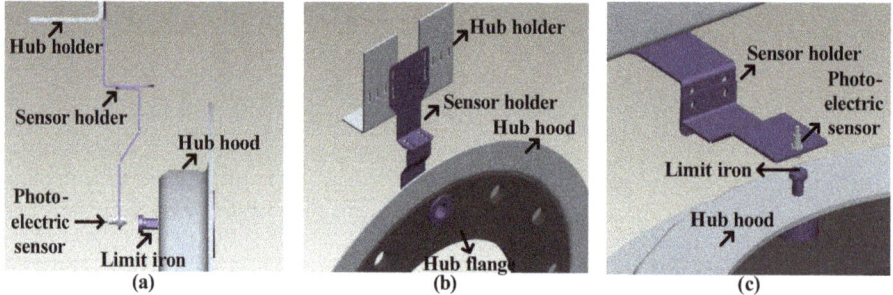

Figure 12. Installation diagram of sensors and mechanical devices. (**a**) is the lateral view, (**b**) is the front view, and (**c**) is a perspective drawing.

Figure 13. Installation diagram of the OPC2 module in the pitch controller cabinet.

The control logic of the OPC2 module is shown in Figure 14. Every sensor provides a speed and a state, which indicates whether or not it is working normally in a sampling period. Three speeds will be transferred as three overspeed states as the inputs in Figure 15 [26].

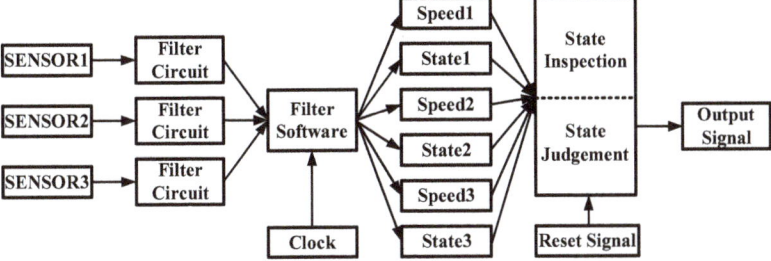

Figure 14. The control logic of the OPC2 module.

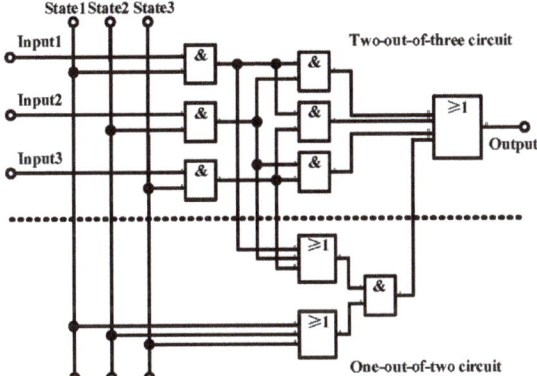

Figure 15. The improved 2-out-of-3 circuit.

An improved 2-out-of-3 circuit is used in the state judgement in Figure 14. If two high levels of the three overspeed states are detected, the module will output a high level (overspeed signal). Furthermore, a 1-out-of-2 circuit is added in the 2-out-of-3 circuit, the control logic of which is if one of the three incorrectly working sensors, and one high level in the two remaining overspeed states will trigger a high level of the module output. If two sensors work incorrectly, the OPC2 will send an overspeed signal directly.

6. Validation and Test Results

The effectiveness validation of the holistic control scheme is based on laboratory tests and field tests. The laboratory tests are used to test the OPC function in Section 5 and validate the effectiveness of the control circuit in Section 4.2.3. The pitching performance of a WT is evaluated during a ZVRT (simulating an overspeed condition on an operating WT is definitely forbidden). The field tests are used to validate the effectiveness of the control logic in Section 4.1 during a ZVRT.

First, a set of rotor speeds is inputted into the OPC2 module. As shown in Figure 16, the rotor speeds continue increasing linearly until reaching 20.35 rpm (1.17 p.u.) in the first 11.5 s, remain unchanged for the next 10 s, and then decrease linearly to 0. The EFC-triggering curve in Figure 16 illustrates that the module can output a correct command (an overspeed signal), interrupt the safe loop in time, and hold the command even if the rotor speeds decrease in the last 11.5 s, which means that the module could execute a required OPC function.

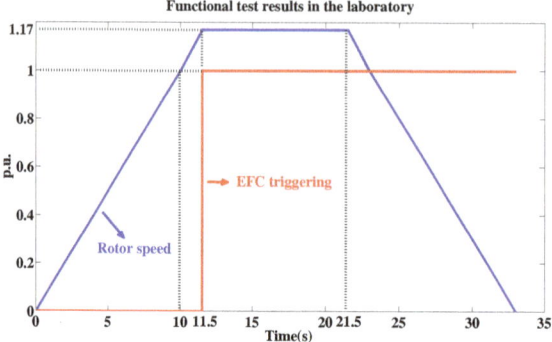

Figure 16. Functional test results of the new OPC module in the laboratory. On the y-axis, a value of 1 for the EFC triggering indicates the high level, and the p.u. value of the rotor speed is 17.4 rpm.

Second, a laboratory test, as shown in Figure 17, is conducted. The laboratory test could simulate the power-loss condition (the three-phase voltages at PCC dip to 0%, approximately). The dragging motor is to simulate the condition (torque) of pitching one blade. The pitch actuator is installed under the pitch controller in the axisbox. The model, which runs on a computer (not shown in Figure 17), simulates the WT controller using the pitch angle control strategy in Section 3.3, and the control logic in Section 4.1.

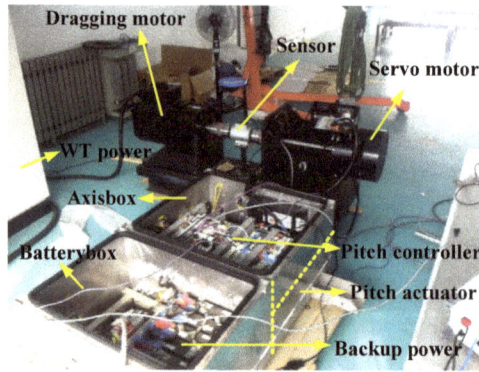

Figure 17. Schematic diagram of the laboratory test.

The test also selects a set of field data (such as the wind speed curve in Figure 18). To meet the field conditions, the threshold value of triggering an overspeed signal in OPC2 module is temporarily reset to 15 rpm. Therefore, a wind speed reaching 0.5 p.u. will trigger an EFC at the moment of 0 s in Figure 18. All pitch angles for the three blades reach 91° in 10 s, proving the effectiveness of the control circuit in Section 4.2.3 and the OPC system. The following performance of the three pitch subsystems is even beyond the design requirement. The reference pitch angle rises from 0° to 72° in the 10 s, which means that the reference pitching speed is nearly 7 deg/s. However, the actual pitching speed can reach nearly 9 deg/s because any pitching speed greater than 7 deg/s will be encouraged in the emergent mode. The WT controller will output a fault only for a value less than 7 deg/s.

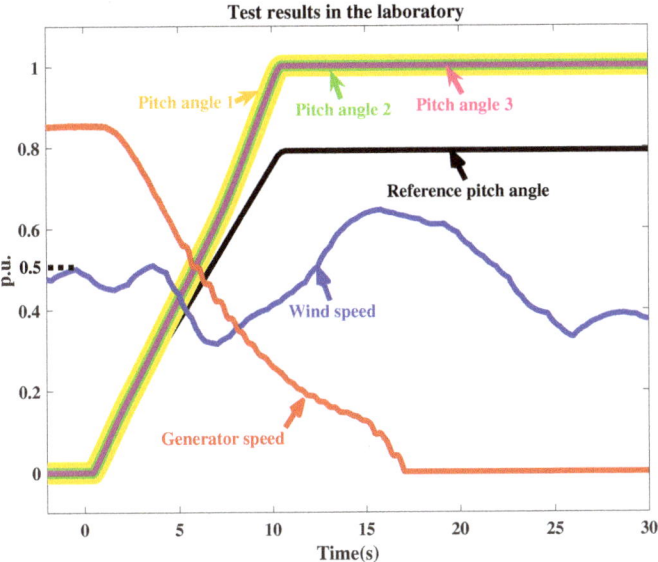

Figure 18. Test results in the laboratory. The p.u. values of the pitch angle, generator speed and wind speed are 91°, 1750 rpm and 11 m/s, respectively.

Finally, an operating WT in Hebei, China, is transformed with the holistic control scheme, and then a field test of ZVRT is conducted according to the test method suggested by IEC [30] (the ZV conditions are simulated according to Figure 19). The ride-through period is defined as 430 ms according to the South Australian standard [32], which is the strictest in the world [12]. The active power of the WT shown in Figure 20 indicates that the WT has ridden through a ZV dip successfully and not triggered the emergent mode during ZVRT. The pitch angle in Figure 20 verifies that the pitch system does execute a feather command at the start of the ZV dip and a fine command at the end of the ZV dip successfully according to the control logic in Section 4.1.

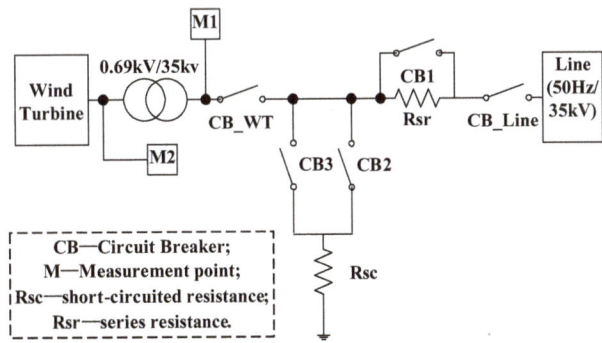

Figure 19. The method of the ZVRT field test.

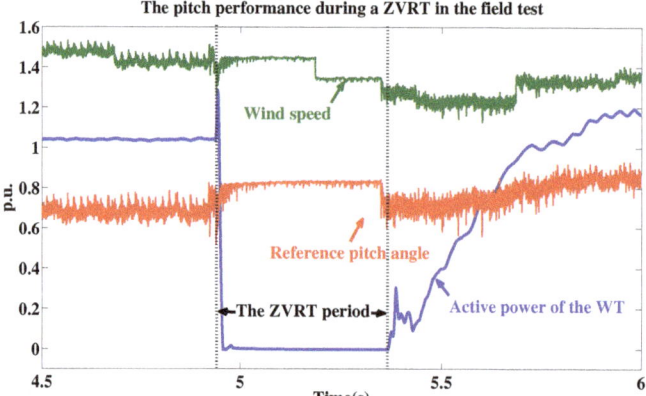

Figure 20. Results of the field test. The p.u. values of the reference pitch angle, active power, and wind speed are 10°, 2 MW and 10 m/s, respectively. The fault condition is a three-phase (symmetrical) voltage dip.

The results of the laboratory and field tests demonstrate the effectiveness of the holistic control scheme and that it satisfies the ZVRT requirements for a WT. The control scheme also has the advantage of contributing to improving the grid-connection capability of WTs, as described in Section 1.

7. Conclusions and Future Work

The early work for this study confirmed that the control scheme with the variable-speed pitch system for WTs during a LVRT cannot satisfy the ZVRT requirements because the WT controllers using this scheme cannot distinguish a ZVRT status from a power-loss condition.

This paper first proposes a holistic control scheme for the pitch system during a ZVRT that mainly includes the control logic programmed into the WT and pitch controllers, some control circuits, and the OPC system. According to the test results, the proposed control scheme perfects the fault response technology and improves the grid-connection performance for WTs. The following additional conclusions can be made:

(1) It is highly important to model the mechanical devices in the aerodynamic and pitch systems to simulate the pitching process. The aerodynamic parameters in Table 1 are used to calculate the torque of pitching the blade, which is the basis of simulating the pitching process. The parameters of the pitch driver (servo motor) in Table 1 should match the torque.

(2) If the blade in Table 1 was replaced with a larger one (the aerodynamic parameters were changed), the pitching torque would be increased. It would be hard for the pitch driver in Table 1 to pitch the larger blade. The pitching speed in test results would be decreased and the pitching time would be longer, which meant a worse pitching performance. Conversely, the pitching speed would be increased in a certain extent and the pitching time would be shorter, which meant a better pitching performance.

(3) Monitoring and protective devices are critical for the primary control circuits in WTs. Control device damage is unavoidable during operation, but incorrect outputs of the primary control circuits can lead to emergencies. Therefore, the stability of the control device should be reconsidered when WTs are operating (or are about to operate) in extreme environments (high-temperature, high-humidity, high-dust, etc).

Some limitations of this study and future work are summarized as follows:

(1) Further observation and research on the adaptability and robustness of the control scheme, supported by more data from long-running systems under various external environments and grid

conditions would be useful. It is suggested that several WTs be equipped with the proposed scheme in an environment to suppress occasional interference.

(2) The emergent-feathering performance of WTs should be tested periodically using the control scheme with a pitch system. An improved test method is provided in the Appendix B.

Author Contributions: Conceptualization, E.C.; methodology, E.C. and Y.Y.; software, E.C.; validation, E.C.; writing—original draft preparation, E.C.; writing—review and editing, L.D. and X.L. All authors have read and agreed to the published version of the manuscript.

Funding: This research received no external funding.

Conflicts of Interest: The authors declare no conflict of interest.

Abbreviations

The following abbreviations are used in this manuscript:

ZVRT	Zero-voltage ride through
LVRT	Low-voltage ride through
FRT	Fault ride through
WT	Wind turbine
OPC	Overspeed protection control
PCC	Point of common coupling
ZV	Zero-voltage
LV	Low-voltage
RSC	Rotor-side converter
GSC	Grid-side converter
DC	Direct current
PI	Proportional integral
PID	Proportional-Integral-Derivative
EFC	Emergency feather command
TOF	Turn-off delayer
UPS	Uninterrupted power supply
NC	Normally closed
NO	Normally open
IEC	International Electrotechnical Commission
IEEE	Institute of Electrical and Electronics Engineers

Appendix A. Standard Grid Codes for LVRT and ZVRT

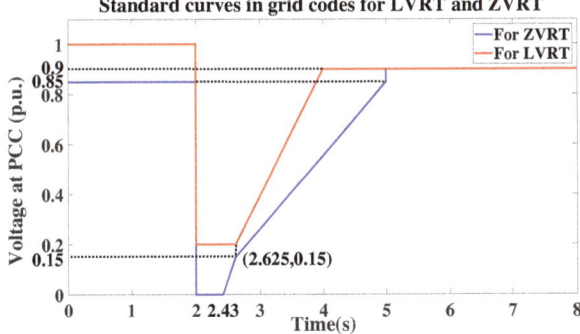

Figure A1. Standard curves in grid codes for LVRT and ZVRT. The LVRT and ZVRT curves shown using the grid codes for China and South Australia [32,36].

Appendix B. Regular Test Method of Emergent-Feathering

It is stipulated in the WT design documents that the emergent-feathering performance must be tested regularly. The period is usually three months or half a year. The common method is to interrupt the safe loop intentionally to simulate an emergency and then to observe the pitch performance.

On 25 December 2018, a 3 MW WT operating in Hebei, China, could not feather during the test until the UPS in the WT controller was cut off in the emergency. After analysis, the accident might be correlated with the design flaws in Section 4.2. The accident process can be summarized as follows. First, two signal slides, the EFC signal in Figure 6 and 24 V UPS, were connected in the slip ring (the slip ring of a WT has limited space but a complicated wiring design and is a mechanical rotating device, and thus breakover errors are typically unavoidable). Then, when the safe loop was interrupted, the WT disconnected from the grid, which meant that the WT lost electromagnetic torque. Finally, driven by the mechanical torque, the aerodynamic system of the WT rotated with an overspeed, and the operating status is shown in Figure A2.

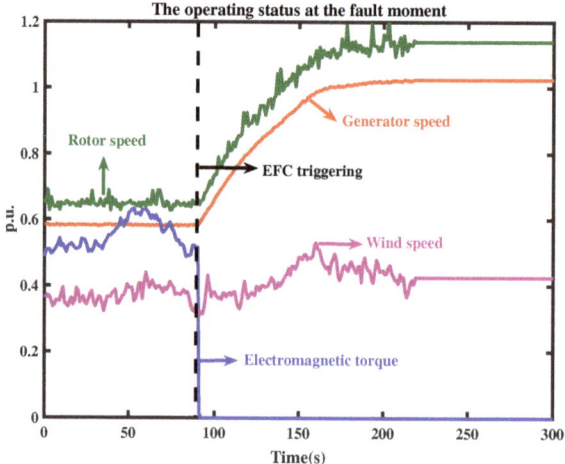

Figure A2. The operating status of the WT at the fault moment. The p.u. values of the electromagnetic torque, generator speed, rotor(hub) speed, and wind speed are 4300 N*m, 1200 rpm, 12.9 rpm, and 10 m/s, respectively.

Figure A2 also shows that the rotor speed will double in approximately 70 s for the conditions of losing the electromagnetic torque and not-pitching, even if the wind speed is very low (nearly 4 m/s).

Overall, the improved test method is suggested as follows on the condition that the wind speed is less than 6 m/s and there is no rain.

(1) Stop the WT, lock the hub with bolts, check the pressure of the hydraulic pressure station, and pitch just one blade towards fine to 0° manually using the WT power.

(2) Press the EFC button in the WT controller cabinet, which also enables the backup power for the pitch system.

(3) Measure the capacitor voltages or battery impedances in the backup power.

(4) Repeat the above steps two times, and draw three feather curves in one plot for three blades (the x-axis represents the feather time, and the y-axis records the pitch angle, such as Figure 18).

The qualified performances can be defined as follows.

(1) Three pitch angles have finally reached 91°.

(2) The entire feather time is less than 15 s. It will be also acceptable that the feather times in two tests are both less than 17 s, and all pitch angles reach 91°. When the air temperature is below −20 °C,

the limit of 17 s can be broadened to 21 s, but the capacitor voltages or battery impedances should meet the requirements.

(3) The feather curves should be smooth, and the following performance should satisfy the requirement that the differences among the pitch angles should be less than 10° at any moment.

References

1. Zhu, D.; Zou, X.; Deng, L.; Huang, Q.; Zhou, S.; Kang, Y. Inductance-emulating control for DFIG-based wind turbine to ride-through grid faults. *IEEE Trans. Power Electron.* **2017**, *32*, 8514–8525. [CrossRef]
2. Chang, Y.; Hu, J.; Tang, W.; Song, G. Fault Current Analysis of Type-3 WTs Considering Sequential Switching of Internal Control and Protection Circuits in Multi Time Scales During LVRT. *IEEE Trans. Power Syst.* **2018**, *33*, 6894–6903. [CrossRef]
3. Zhang, Y.; Melin, A.; Djouadi, S.; Olama, M.M.; Tomsovic, K. Provision for Guaranteed Inertial Response in Diesel-Wind Systems via Model Reference Control. *IEEE Trans. Power Syst.* **2018**, *33*, 6557–6568. [CrossRef]
4. Liu, R.; Yao, J.; Wang, X.; Sun, P.; Pei, J.; Hu, J. Dynamic Stability Analysis and Improved LVRT Schemes of DFIG-Based Wind Turbines During a Symmetrical Fault in a Weak Grid. *IEEE Trans. Power Electron.* **2020**, *35*, 303–318. [CrossRef]
5. Yang, L.; Xu, Z.; Østergaard, J.; Dong, Z.Y.; Wong, K.P. Advanced control strategy of DFIG wind turbines for power system fault ride through. *IEEE Trans. Power Syst.* **2012**, *27*, 713–722. [CrossRef]
6. Shen, Y.; Ke, D.; Qiao, W.; Sun, Y.Z.; Kirschen, D.S.; Wei, C. Transient Reconfiguration and Coordinated Control for Power Converters to Enhance the LVRT of a DFIG Wind Turbine With an Energy Storage Device. *IEEE Trans. Energy Convers.* **2015**, *30*, 1679–1690. [CrossRef]
7. Yang, S.; Zhou, T.; Chang, L.; Xie, Z.; Zhang, X. Analytical method for DFIG transients during voltage dips. *IEEE Trans. Power Electron.* **2017**, *32*, 6863–6881. [CrossRef]
8. Guo, W.; Xiao, L.; Dai, S.; Li, Y.; Xu, X.; Zhou, W.; Li, L. LVRT Capability Enhancement of DFIG With Switch-Type Fault Current Limiter. *IEEE Trans. Ind. Electron.* **2015**, *62*, 332–342. [CrossRef]
9. Cheng, M. Research on the Key Technologies of Low Voltage Ride through for Doubly-Fed Wind Power Generation System. Ph.D. Thesis, Shanghai Jiao Tong University, Shanghai, China, 2012.
10. Ouyang, J.; Xiong, X. *Electro-Magnetic Transient Analysis in Doubly-Fed Wind Generation Systems*, 1st ed.; Science Press: Beijing, China, 2018.
11. Zhu, D., Zou, X.; Zhou, S.; Dong, W.; Kang, Y.; Hu, J. Feedforward current references control for DFIG-based wind turbine to improve transient control performance during grid faults. *IEEE Trans. Energy Convers.* **2018**, *33*, 670–681. [CrossRef]
12. Zhao, H. The Research of Zero Voltage Ride Through Strategy of DFIG. In Proceedings of the IET International Conference on Renewable Power Generation (RPG 2016), London, UK, 21–23 September 2016.
13. Zhao, H.; Tang, H.; Zhang, W.; Wen, W.L. Transient Characteristics Research and Integrated Control Strategy of DFIG for Zero Voltage Ride Through. *Power Syst. Technol.* **2016**, *40*, 1422–1430.
14. Tang, H.; Chang, Y.; Chi, Y.; Wang, B.; Li, Y.; Hu, J. Analysis and Control of Doubly Fed Induction Generator for Zero Voltage Ride Through. In Proceedings of the 19th International Conference on Electrical Machines and Systems (ICEMS), Chiba, Japan, 13–16 November 2016.
15. Cai, E.; Jiao, C.; Wang, D.; Jing, J.; Huang, F.; Pan, F. Test scheme of zero voltage ride through capability for DFIG-based wind turbines. *Autom. Electr. Power Syst.* **2016**, *40*, 137–142.
16. Zhang, C. Research on Individual Variable Pitch Control Strategies of Large-Scale Wind Turbine. Ph.D. Thesis, Shenyang University of Technology, Shenyang, China, 2011.
17. Dou, Z. Research on Individual Pitch Control for Large-Scale Wind Turbine. Ph.D. Thesis, Shanghai Jiao Tong University, Shanghai, China, 2013.
18. Burton, T.; Jenkins, N.; Sharpe, D.; Bossanyi, E. *Wind Energy Handbook*, 2nd ed.; John Wiley & Sons: Chichester, UK, 2011.
19. Zhang, Y.; Melin, A.; Djouadi, S.; Olama, M. Performance guaranteed inertia emulation for diesel-wind system feed microgrid via model reference control. In Proceedings of the 2017 IEEE Power&Energy Society Innovative Smart Grid Technologies Conference(IGST), Washington, DC, USA, 23–26 April 2017.
20. Van, T.; Nguyen, T.; Lee, D. Advanced pitch angle control based on fuzzy logic for variable-speed wind turbine systems. *IEEE Trans. Energy Convers.* **2015**, *30*, 579–587. [CrossRef]

21. Jafamejadsani, H.; Pieper, J. Gain-Scheduled ℓ_1-Optimal Control of Variable-Speed-Variable-Pitch Wind Turbines. *IEEE Trans. Contr. Syst. Technol.* **2015**, *23*, 372–379. [CrossRef]
22. Chen, P.; Han, D.; Tan F.; Wang, J. Reinforcement-Based Robust Variable Pitch Control of Wind Turbines. *IEEE Access* **2020**, *8*, 20493–20502. [CrossRef]
23. Tang, X.; Yin, M.; Shen, C.; Xu, Y.; Dong, Z.Y.; Zou, Y. Active Power Control of Wind Turbine Generators via Coordinated Rotor Speed and Pitch Angle Regulation. *IEEE Trans. Sustain. Energy* **2019**, *10*, 822–832. [CrossRef]
24. Chirca, M.; Dranca, M.; Teodosescu, P.; Breban, S. Limited-Angle Electromechanical Actuator for Micro Wind Turbines Overspeed Protection. In Proceedings of the 2019 11th International Symposium on Advanced Topics in Electrical Engineering (ATEE), Bucharest, Romania, 28–30 March 2019.
25. Wen, L.; Sheng, W.; Xu, Z. Research on overspeed protection control strategies of PWR nuclear power units. In Proceedings of the 2019 IEEE Innovative Smart Grid Technologies-Asia (ISGT Asia), Chengdu, China, 21–24 May 2019.
26. Ji, J. On Fail-safety Control Technology of 1000 MW Steam Turbine for Nuclear Power Plants and its Application. Master's Thesis, Shanghai Jiao Tong University, Shanghai, China, 2018.
27. Wang, L.; Zhao, J.; Liu, D.; Wang, J.; Chen, G.; Sun, W.; Qi, X. Governor tuning and digital deflector control of Pelton turbine with multiple needles for power system studies. *IET Gener. Transm. Distrib.* **2017**, *11*, 3278–3286. [CrossRef]
28. Mellal, M.; Chebouba, B. Cost and Availability optimization of Overspeed Protection System in a Power plant. In Proceedings of the 2019 International Conference on Advanced Electrical Engineering (ICAEE), Algiers, Algeria, 19–21 November 2019.
29. IEEE 1865.1-2019. *IEEE Standard Specifications for Maintenance and Test of Distributed Control Systems in Thermal Power Stations: Maintenance and Testing*; IEEE: Piscataway Township, NJ, USA, 2019.
30. IEC 61400-21-1:2019. *Wind Energy Generation Systems—Part 21-1: Measurement and Assessment of Electrical Characteristics—Wind Turbines*; IEC: Geneva, Switzerland, 2019.
31. IEEE 1159-2019. *IEEE Recommended Practice for Monitoring Electric Power Quality*; IEEE: Piscataway Township, NJ, USA, 2019.
32. South Australian Minister. *National Electricity Rules, Version 142*; AEMC: Sydney, Australia, 2020.
33. Northern Territory Minister. *National Electricity Rules as in Force in the Northern Territory, Version 52*; AEMC: Sydney, Australia, 2020.
34. Module Introduction–KL2134. Available online: https://www.beckhoff.com/english/bus_terminal/kl2134.htm (accessed on 19 July 2017).
35. Module Introduction–KL1104. Available online: https://www.beckhoff.com/english/bus_terminal/kl1104.htm (accessed on 19 July 2017).
36. GB/T 19963. *Technical Rule for Connecting Wind Farm to Power System*; SAC: Beijing, China, 2016.

© 2020 by the authors. Licensee MDPI, Basel, Switzerland. This article is an open access article distributed under the terms and conditions of the Creative Commons Attribution (CC BY) license (http://creativecommons.org/licenses/by/4.0/).

Article

A Hybrid RCS Reduction Method for Wind Turbines

Shyh-Kuang Ueng

Department of Computer Science & Engineering, National Taiwan Ocean University, Keelung 202, Taiwan; skueng@mail.ntou.edu.tw

Received: 26 June 2020; Accepted: 25 September 2020; Published: 29 September 2020

Abstract: Wind turbine towers produce significant scatterings when illuminated by radars. Their reflectivity affects air traffic control, military surveillance, vessel tracking, and weather data sensing processes. Reducing the radar cross-section (RCS) of wind turbines is an essential task when building wind farms. It has been proved that round and bumpy structures can scatter radar waves and reduce the RCS of a reflector. Other research showed that taper towers generate smaller radar returns than cylindrical towers. In this research, we combine both strategies to devise a more effective method for designing wind turbine towers in the hope that their RCS can be further reduced. The test results reveal that the proposed method out-performs current reshaping methods. Wind turbine towers possessing taper shapes and periodic surface bumps deflect incident electromagnetic waves to insignificant directions. Thus, radar returns in the back-scattering directions decrease. Other experiments also verify that the proposed method maintains its effectiveness for radar waves with varying frequencies and polarization.

Keywords: RCS reduction; wind turbines; reshaping methods

1. Introduction

The massive metallic towers of wind turbines produce significant back scatterings when illuminated by radar waves. Their radar cross-sections (RCS) are larger than those of Boeing 737 airplanes. Thus, wind turbines become a menace to the operations of air and sea traffic control, weather monitoring, and military surveillance radars [1]. This hazard is the top reason for the cancellation of wind farm installations [2]. The study in [2] surveyed the scattering capabilities of all individual components of a wind turbine and concluded that 75% of the radar returns were produced by the tower. If the radiation from the tower is decreased, the entire scattering of the wind turbine is reduced too. Furthermore, the tower is used for supporting the electricity generator, blades, and nacelle. It is a stationary structure and does not rotate as the blades do. Applying RCS alleviation processes upon the tower is more economic and produces almost no negative effects on the electricity generation capacity of the wind turbine. Thus, alleviating radar returns from the tower is the top priority for decreasing the RCS of a wind turbine.

Some strategies have been proposed to reduce the radar returns caused by wind turbine towers. In the work of [3,4], researchers proposed to coat these metallic structures with radar absorption materials (RAM) to weaken their scatterings. RAM methods possess some engineering difficulties. First, the thickness of the coating layer must be properly calculated so that radar waves of a specific frequency can be absorbed. Secondly, RAMs are expensive, and their installation costs are high. Third, their weights increase the load of the tower. Furthermore, the endurance of RAMs is another problem since some wind turbines are installed off-shore and erosion can damage the coatings.

An alternative methodology is to change the shape of the tower to divert incident radar waves toward insignificant directions so that the back scatterings are declined. Based on this rationale, some reshaping methods have been proposed. In the paper of [2], Pinto et al. proposed to shape a wind turbine tower with tapering effect. The resultant tower has a larger base end and a smaller top end.

It has been proved that this tapered tower produces less radar returns than a cylindrical tower when illuminated by S- and X-band radars. In [5], Ling et al. discovered that adding bump structures on the surface of a reflector could effectively divert the reflected EM waves. Their approach was revised by Ueng et al. to lower the RCS of wind turbines [6,7]. In the methods of [6,7], the surfaces of wind turbine towers are augmented with horizontal or vertical periodic bumps to scatter incident radar waves. Hence, radar returns from wind turbine towers are alleviated. The authors also published and analyzed simulation results and verified that bumps in specific densities and heights significantly reduce RCS of wind turbine towers. In a more recent work [8], Ueng and Chen proposed to combine the two reshaping strategies of [2,6,7] to design a more efficient RCS reduction method. However, a pragmatic algorithm for building wind turbine towers was absent from their paper.

Besides reshaping and RAM methods, some researchers proposed to use metasurfaces for RCS reduction. In their methods, layers of metallic and dielectric slabs are adhered in specific arrangements. The thicknesses and covering areas of these slabs are carefully selected to produce out-of-phase effects in the reflected radar waves. Thus, the RCS of the surface is reduced because of the destructive interferences between reflected beams [9]. In a similar search, Song et al. designed a hybrid RCS alleviation method [10]. They created a graphene sheet upon the target surface. The space between the graphene sheet and the target surface was filled with foams. Then, they attached grating structures on the graphene sheet. As radar waves reach the target, the graphene layer absorbs some energy from the incoming waves and the grating structures generate high order reflections to reduce the back scatterings. In the work of [11,12], scientists employed artificial magnetic conductors (AMC) to weaken the RCS of metallic surfaces. In these methods, specially designed AMC cells are designed and fabricated on metallic surfaces. These AMC cells form a chessboard pattern. As metallic surfaces are illuminated by radar waves, the ACM cells diversify the phase distribution of the reflected beams and generate phase-cancellation effects. Thus, the RCS of the target is decreased. These methods do not modify the shape of the target. However, they share some weaknesses with RAM methods. Their costs are relatively high and the endurance of the coating layers in harsh environments is questionable.

In this research, we adopt the idea of [8] and develop an innovative reshaping method aiming to design wind turbine towers producing a smaller RCS. Our method combines the tapering effect method of [2] and the bump surface methods of [5–7] to model wind turbine towers. The resultant towers possess tapered shapes and periodic surface bumps. Besides devising and formulating our reshaping approach, numerous simulations have also been carried out to study the effectiveness of the proposed method. The collected results show that our hybrid reshaping method is superior to those previous reshaping methods presented in [2,5–7]. Furthermore, the achieved RCS reduction exceeds the combination of the individual reshaping methods. Other experiments show that the proposed strategy is applicable for a wide range of frequencies and different radar wave polarizations. Thus, it can be used to decrease scatterings for various radar systems. We also conduct simulations to find the optimal tapering angles and to study the combined effect of taper ratio and radar frequency upon the proposed reshaping strategy. The results are also presented and analyzed in this article.

2. Materials and Methods

As the proposed RCS reduction method is a hybrid reshaping strategy, the resultant towers possess tapered shapes and surface bumps. There are fine structures on the towers' surfaces, and the taper ratio of each individual tower may be different. Instead of designing these towers manually by using geometric modelling tools, we deduce mathematic formulas to model them. The bump density, bump height, bump direction, and taper ratio are regarded as control parameters in the formulas. Thus, by altering these variables, towers with different appearances can be automatically created to comply with the requirements of the users.

2.1. Cylindrical, Taper, Bump, and Bamboo Tower Modelling

The reshaping methods for modelling towers with bumpy and tapering effects have been presented in [5–7]. Based on these reshaping methods, we developed new geometrical formulations to create hybrid towers. For the sake of completeness, the reshaping methods of [5–7] will be first introduced in this subsection. The newly developed modeling algorithms for the hybrid towers is presented in Section 2.2.

Our reshaping algorithm uses a cylinder tower as the template. Then, by deforming the cylindrical tower, other types of towers are constructed. In this subsection, we will present the proposed modelling formula step by step and case by case. We assume that, in the world coordinate system, the x- and y-axes span the horizontal plane while the z-axis points vertically, and the height and radius of the cylindrical tower are h and r. Then, the coordinates of all points on the cylindrical tower surface can generated by the following formula:

$$\begin{aligned} x &= r\cos(\alpha), \\ y &= r\sin(\alpha), \\ \frac{-h}{2} &\le z \le \frac{h}{2}, \end{aligned} \quad (1)$$

where α ranges from 0 to 2π. An image containing a cylindrical tower and the world coordinate system is shown in Figure 1a.

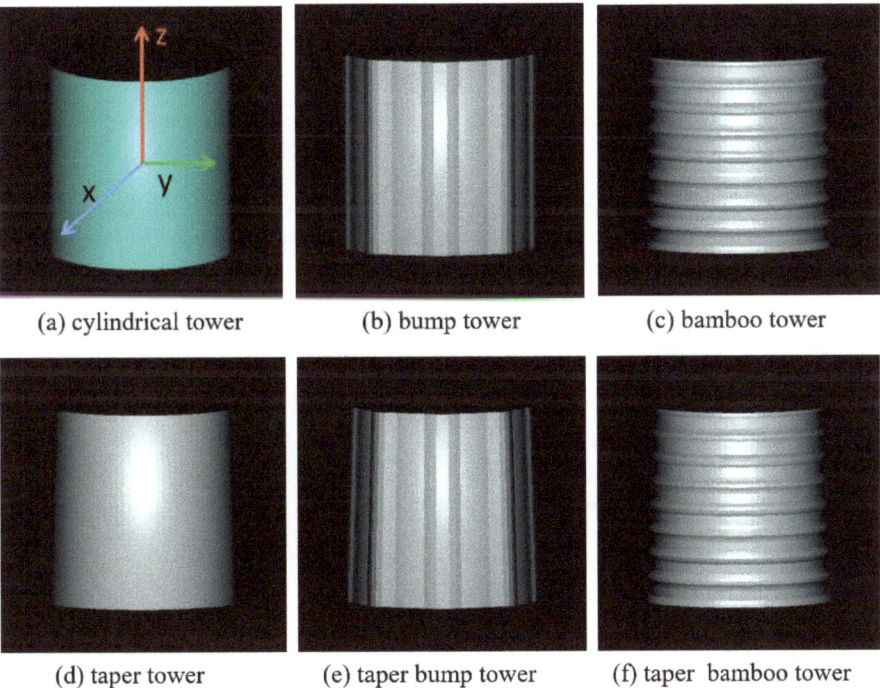

Figure 1. Wind turbine towers. Upper row, towers without tapering effect; lower row: towers with tapering effect.

Based on Equation (1), we deduced a method to construct a tapered tower, assuming that the radii of the bottom end and the top end of the tapered tower are r_B and r_T. The slope, i.e., the taper ratio, of the tower surface is defined as:

$$T = \frac{r_B - r_T}{h}. \quad (2)$$

The radius of the tapered tower varies with the z-coordinate and can be expressed as a function of z-coordinate:

$$r(z) = r_B - T * (z + \frac{h}{2}). \tag{3}$$

By substituting $r(z)$ of Equation (3) into Equation (1), we can generate a tapered tower with a taper ratio T. An example taper tower is displayed in Figure 1d.

By adding perturbations to the x- and y-coordinates in Equation (1), we can generate bumps on a cylindrical tower's surface. These perturbations come from a wave function $D(\alpha)$, which is defined as follows:

$$D(\alpha) = A \cos(k\alpha), \tag{4}$$

where A and k are the amplitude and wave number of this wave function, respectively, and α ranges from 0 to 2π. Then, the x- and y-coordinates of the tower surface can be modelled by using the following equation:

$$\begin{aligned} x(z,\alpha) &= (D(\alpha) + r) \cos(\alpha), \\ y(z,\alpha) &= (D(\alpha) + r) \sin(\alpha). \end{aligned} \tag{5}$$

In Equation (5), the perturbation function alters the radius of the tower and transforms the tower surface into a wavy surface. The crests divert incoming radar rays and reduce the RCS of the tower. However, the troughs resemble parabolic reflectors. They may concentrate incident radar rays and produce significant back scatterings. To preserve the crests and eliminate the troughs, we put the following constraint upon $D(\alpha)$:

$$D(\alpha) = \max(A \cos(k\alpha), 0). \tag{6}$$

Subsequently, D becomes a non-negative periodic function of α and generates a sequence of convex bumps around the tower surface. The tower is named as the bump tower in this article. A bump tower is shown in Figure 1b. The bumps are parallel to the z-direction.

By revising the periodic function D, we can reorient the bumps by 90 degrees and add equal-spaced rings on the tower surface. Assuming that k rings are to be produced, the inter-ring space is computed by

$$\lambda = \frac{h}{k}. \tag{7}$$

By using λ, we redefine D as a function of the z-coordinate:

$$D(z) = \max(A \cos(\frac{2\pi z}{\lambda}), 0). \tag{8}$$

Consequently, D becomes a periodic and non-negative function of z-coordinate. By substituting D of Equation (8) into Equation (5), the x- and y-coordinates of the tower surface are computed as below:

$$\begin{aligned} x(z,\alpha) &= (D(z) + r) \cos(\alpha), \\ y(z,\alpha) &= (D(z) + r) \sin(\alpha). \end{aligned} \tag{9}$$

Since the tower possesses a series of convex rings and resembles a bamboo, we call it the bamboo tower in this article. The image of a bamboo tower is contained in Figure 1c. Its surface contains horizontal rings which are expected to reflect incident radar waves to the ground and the sky.

2.2. Hybrid Tower Modelling

It has been proved that towers possessing surface bumps or taper shapes produce less RCS than cylindrical towers [2,5–7]. A reshaping method which combines both strategies would be of great values to us. However, to our knowledge, no study has been carried out to investigate the RCS reduction capability using both methods at the same time. Thus, we created two types of hybrid towers which have surface bumps and tapering shapes in hope to further alleviate back scatterings from wind turbine towers.

First, we mixed the bump tower with the taper tower to create a taper bump tower. To achieve this goal, we deduced the following function to control the radius r of the tower:

$$r(\alpha, z) = (r_B + D(\alpha)) - T * (z + \tfrac{h}{2}),$$
$$-h/2 \leq z \leq h/2, \qquad (10)$$
$$0 \leq \alpha \leq 2\pi,$$

where T and $D(\alpha)$ are defined in Equations (2) and (6), respectively. Thus, the radius is maximized at the base and linearly decreases along with the z-coordinate. Furthermore, the radius is disturbed by the periodic function D in each cross-section of the tower to create convex bumps. By using the radius function $r(\alpha, z)$, the x- and y-coordinates of the tower surface are modelled by

$$x(\alpha, z) = r(\alpha, z) * \cos(\alpha),$$
$$y(\alpha, z) = r(\alpha, z) * \sin(\alpha). \qquad (11)$$

The image of a taper bump tower is displayed in Figure 1e. As the image shows, the tower has a taper shape and a sequence of bumps surrounding its surface.

Then, by using Equations (7) and (8), we vary the radius along with the z-coordinate

$$r(\alpha, z) = (r_B + D(z)) - T * (z + \tfrac{h}{2}),$$
$$-h/2 \leq z \leq h/2, \qquad (12)$$
$$0 \leq \alpha \leq 2\pi,$$

The radius fluctuates with the z-coordinate and linearly shrinks as the z-coordinate increases. Then, the x- and y-coordinates of the tower surface are calculated by:

$$x(\alpha, z) = r(\alpha, z) * \cos(\alpha),$$
$$y(\alpha, z) = r(\alpha, z) * \sin(\alpha). \qquad (13)$$

The resultant tower is called the taper bamboo tower in the following context. An image of the tapered bamboo tower is displayed in Figure 1f. This tower has a tapering shape as well as horizontal convex rings on its surface.

3. Results

Several sets of experiments were conducted to evaluate the proposed reshaping methods. In the first set of tests, we computed the RCS values of the wind turbine towers mentioned in the previous section. Then, we analyzed the test results to find out effective tower shapes. In the second set of experiments, we studied how taper ratio and radar frequency affected the RCS of these towers. The efficiencies of these towers, when illuminated by radar waves of various frequencies, were also compared. In the third set of simulations, we polarized the radar waves both vertically and horizontally and then computed the RCS of these towers. These simulations reveal the influence of polarization upon the performances of the proposed reshaping methods.

In order to save costs, all the experiments were carried out using a simulation program. We used the geometric modelling methods presented in Section 2 to construct virtual wind turbine towers, including a cylindrical tower, a bump tower, a bamboo tower, a taper tower, a taper bump tower, and a taper bamboo tower. The origin of the world coordinate system is located at the tower centers, as shown in Figure 1a. To speed up the computations, the tower height was truncated to six meters. Initial geometrical parameters of the reshaping procedures are listed in Table 1. The radii of the towers were 1.5 m. The taper ratios of the taper tower, the taper bump tower, and the bamboo tower are 0.4/60. Eight bumps were created on the surfaces of the bump tower, the bamboo tower, and the two hybrid towers. The bump height was 0.1 m. These data represent default values of the geometric parameters but may be changed in experiments to enhance specific effects and tower characteristics.

Table 1. Geometric parameters of the towers.

Base Radius	Number of Bumps	Bump Height	Taper Ratio
1.5 m	8	0.1 m	0.4/60

Unlike RAM methods, the total energy of scattering is not reduced in reshaping methods. The RCS of a target declines because most of the reflected rays are guided to unimportant directions and less energy is sent back toward the transmitting antenna. Evaluating the effectiveness of towers based on their mono-static RCS may result in missing key features of individual reshaping strategies. In this research, we used bi-static RCS to manifest the diversification of energy caused by the tapering effect and bumps. Thus, the characteristics of the towers were better revealed. Furthermore, scatterings in all directions may cause multiple interactions among nearby wind turbines or other objects and interfere radar operations. If the bi-static RCS is decreased, multiple interactions can be alleviated. Therefore, we rely on bi-static RCS to analyze the performances of towers.

3.1. Effectiveness of Hybrid Towers

In the first set of experiments, bi-static RCS values of the hybrid towers were computed. The fundamental radar parameters are depicted in Table 2. The radar was located at the x-axis and was 3000 m away from the origin. Hence, the zenith angle θ of the incident radar waves was 90 degrees while the azimuth angle ϕ of the incident radar waves was zero degrees. One hundred and eighty-one receivers were used to sense scatterings from towers. These receivers were evenly distributed in the boundary of a semicircle. This semicircle was located on the xy-plane and has a radius of 3000 m. The zenith and azimuth angles of these receivers were $\theta = 90°$ and $-90° \leq \phi \leq 90°$. A computer program based on the shooting and bouncing rays (SBR) method [5,13] was employed to compute the RCS of these towers. In the evaluation process, the RCS of the cylinder tower was served as the baseline to verify the efficiencies of the hybrid towers.

Table 2. Radar parameters.

Frequency	Polarization	Radar Distance	Incident Angle
3.0 GHz	H/H	3000 m	$\theta = 90°$, $\phi = 0°$ (x-axis)

3.1.1. Effectiveness of the Tapered Bump Tower

The RCS values of the cylinder tower, the bump tower, the taper tower, and the taper bump tower are displayed in Figure 2. The RCS of the cylinder tower is rendered in green color and used as the baseline. The RCS values of the bump tower, taper tower, and the taper bump tower are shaded in red color. In the left part, the bi-static RCS values of the cylinder tower and the bump tower are depicted. The bump tower produced less RCS around the backscattering directions ($\theta = 90°$, $-20° \leq \phi \leq 20°$). The difference is about 5 dB. In the middle part, the RCS values of the cylinder tower and the taper tower are rendered. The taper tower reduced the radar returns by 10 dB around the backscattering directions. The RCS of the taper bump tower is shown in the right part. This hybrid tower reduced the RCS value by about 20 dB in the backscattering directions. The results show that the taper bump tower is superior to the bump tower and the taper tower. It produces additive improvement. The reduction in RCS gained by the taper bump tower exceeds the sum of the RCS reductions contributed by the taper tower and the bump tower.

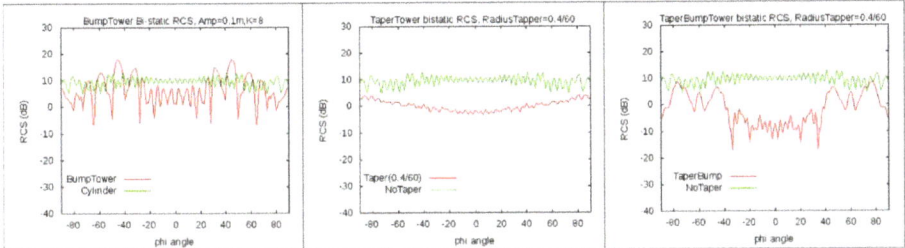

Figure 2. (**Left**) radar cross-section (RCS) of the bump tower. (**Middle**) RCS of the taper tower. (**Right**) RCS of the taper bump tower. RCS of the cylinder tower is rendered in green color and used as the baseline.

The bump tower and taper bump tower generated stronger scatterings around the directions $-70° \leq \phi \leq -50°$ and $50° \leq \phi \leq 70°$ and deteriorated their performances. The incident angle of the radar waves is $\phi = 0°$. After hitting the bumps, their reflections concentrate within these two ranges. Thus, the bistatic RCS at these directions is worsened.

3.1.2. Effectiveness of the Tapered Bamboo Tower

The bi-static RCS values of the taper bamboo tower are displayed in the right part of Figure 3. For reference, the bi-static RCS values of the bamboo tower and taper tower are shown in the left and middle parts of the same figure. The bamboo tower and the taper could reduce the back scatterings by 10 dB. However, the taper bamboo tower reduced the RCS value by 20 dB. Thus, the hybrid reshaping method is superior to the other two towers.

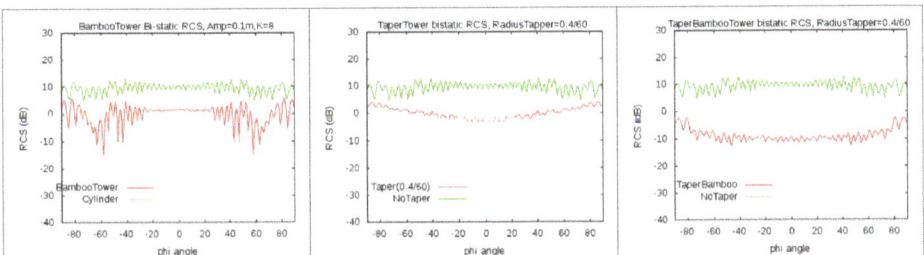

Figure 3. (**Left**) RCS of the bamboo tower. (**Middle**) RCS of the taper tower. (**Right**) RCS of the taper bamboo tower.

Comparing the RCS of the taper bump tower and the taper bamboo tower, we find that the taper bamboo tower can alleviate scatterings in all the azimuth angles. This phenomenon also appears in the RCS of the bamboo tower. We believe that the taper bamboo tower inherits this property from the bamboo tower. Its horizontal bumps scatter the incident radar waves toward the sky and the ground and decrease the bi-static RCS in all azimuth angles.

3.2. Taper Ratio and RCS

If we fix the radius of the tower base, the shape of a tower is decided by the bump height, bump density, bump direction (vertical or horizontal), and taper ratio. Thus, there are four parameters for modelling a tower. The influence of the bump height and bump density on RCS has been reported in the previous research of [6]. The results showed that a feasible bump height for bump towers should be within 0.1 and 0.9 of the radar wavelength. On the other hand, the most effective bump height for bamboo towers is about 0.2~1.4 times that of the radar wavelength. If the bumps are too short, they cannot effectively scatter the incident radar waves. However, if the bumps are too high, the bump bases and the tower surface form corners, which produce significant back-scatterings and enhance RCS.

The reasonable number of bumps is usually between four and eight. Bumps reflect radar waves toward insignificant directions. As the number of bumps increases, the radar returns are gradually weakened. Nonetheless, if the bump density is too high, the space between two bumps resembles a concave reflector and generates a strong reflection toward the receivers, thus, RCS increases. Extra experiment results, analysis, and explanation have been presented in the paper of [6]. Thus, these two factors, bump height and density, will not be investigated in this work. Instead, this research focuses on the influences of taper ratio and bump direction on RCS alleviation. High frequency electromagnetic waves carry higher energies and enhance scatterings [7,9]. Furthermore, their wavelengths are shorter and so they can interact with fine structures on tower surfaces more effectively and produce complicated reflections. Thus, the relationship between radar frequency and RCS will also be studied in the tests.

In preparation for the tests, we varied the taper ratio to create towers. The radar distance, polarization, and direction were not altered, though the radar frequency was varied in the simulations. The RCS of each tower was calculated by using the SBR program mentioned above. In each simulation, this program computed bi-static RCS values in all 181 azimuth directions. To emphasize the backscattering strength, we computed the average value of the bi-static RCS within the range of $-20° \leq \phi \leq 20°$ and used it to evaluate the performances of the towers. Though we did not take RCS in all directions into account, this range of azimuth angles covers the major back scatterings sensed by a mono-static radar. Thus, the average radar return of this range is a reasonable measurement for comparing the efficiencies of the towers.

3.2.1. Fixed Radar Frequency

To uncover the relation between taper ratio and RCS, we created five taper towers with taper ratios of 0.2/60, 0.4/60, 0.6/60, 0.8/60, and 1.0/60. Then, we computed their RCS by using the SBR program. The settings of the simulation were the same as those tests in Section 3.1. The results are illustrated in Figure 4. The green curves represent the RCS of the cylinder tower while the red curves show the RCS of these tapered towers. By examining these five results, we found that the higher the taper ratio, the lower the RCS. Therefore, to alleviate radar returns, we should build towers with high taper ratios.

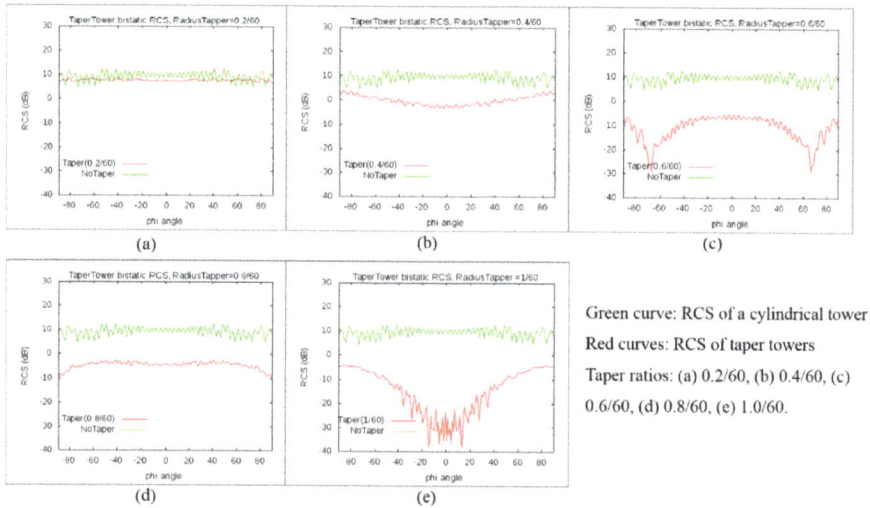

Green curve: RCS of a cylindrical tower
Red curves: RCS of taper towers
Taper ratios: (a) 0.2/60, (b) 0.4/60, (c) 0.6/60, (d) 0.8/60, (e) 1.0/60.

Figure 4. RCS of taper towers with (a) 0.2/60, (b) 0.4/60, (c) 0.6/60, (d) 0.8/60, and (e) 1.0/60 taper ratios.

However, if the taper ratio is 1/60, the difference of diameter between the top and base ends in a 60-m-tall tower is two meters. If the diameter of the top end is three meters, the diameter of the base end will be five meters. Fixing the size of the top end and increasing the taper ratio will widen the difference and the taper tower will have a very large base. In the aspect of building costs, this tower will become impractical. Thus, we suggest that the taper ratio should not exceed 1/60.

3.2.2. Varied Radar Frequency

In following tests, we studied the influence of radar frequency on the RCS of towers. At first, we created six taper towers, six taper bump towers, and six taper bamboo towers. The taper ratios of these towers were 0.0/60, 0.7/60, 1.1/60, 1.3/60, and 1.5/60, respectively. Then, we illuminated each tower by using radar waves of six different radar frequencies (1~6 GHz). The average bi-static RCS of the directions, $\theta = 90°$ and $-20° \leq \phi \leq 20°$, were calculated and used as the metrics for comparisons.

The results are illustrated in Figure 5. The RCS of towers with a 0.0/60 taper ratio is represented by red curves. The RCS of other towers are rendered in green, blue, purple, cyan, and brown colors. In part (a), the test results of the taper towers are displayed. As the image shows, the RCS of the taper tower with a 0.0/60 taper ratio increased along with the radar frequency. On the other hand, the RCS values of other taper towers slightly decreased as the radar frequency increases. Higher frequency radar waves are more similar to light rays than lower frequency radar waves. Scatterings caused by the tapering surfaces become important, and thus more energy of the incident electromagnetic waves is reflected toward insignificant directions.

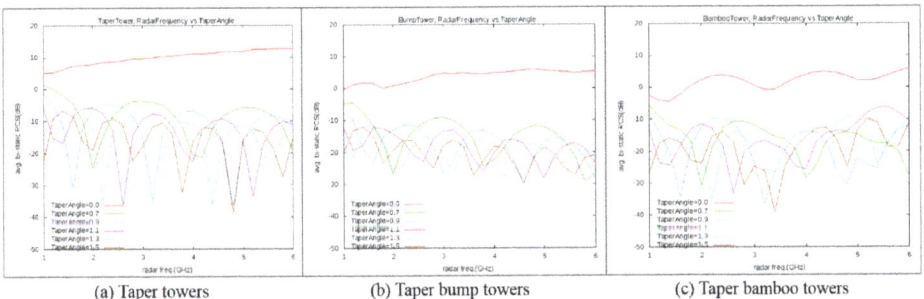

Figure 5. Average bi-static RCS of (**a**) taper towers, (**b**) taper bump towers, and (**c**) taper bamboo towers. Radar frequency: 1~6 GHz, taper ratios: 0.0/60~1.5/60.

The simulation results reveal that taper bump towers and taper towers are effective in reducing RCS. However, taper bump towers generate a lower RCS than taper towers. The bumps on their surfaces help to scatter more energies of the incident electromagnetic waves. Hence, their RCS is further decreased. The RCS of the taper bamboo towers are presented in part (c) of Figure 5. As the radar frequency increases from 1 GHz to 3 GHz, their RCS values decrease. Then, their RCS magnitudes increase as the radar frequency increases. This phenomenon is more obvious for taper bamboo towers with higher taper ratios. In the taper bamboo towers, the bumps are horizontal rings. As the taper ratio increases, the reflections from one ring will be further reflected by neighboring rings and cannot be directed to the sky or the ground straightly. Thus, more energy is sensed by the receivers. As the radar frequency exceeds 4 GHz, these scatterings become more important and the RCS of the bamboo towers are enlarged.

3.3. Tower Type and RCS

In another study, we compared the performances of taper towers, taper bump towers, and taper bamboo towers. The simulation settings and measurement metrics were the same as the previous experiments. The results are depicted in Figure 6. The RCS of the taper towers, the taper bump towers, and the taper bamboo towers are represented by the red, green, and blue curves, respectively.

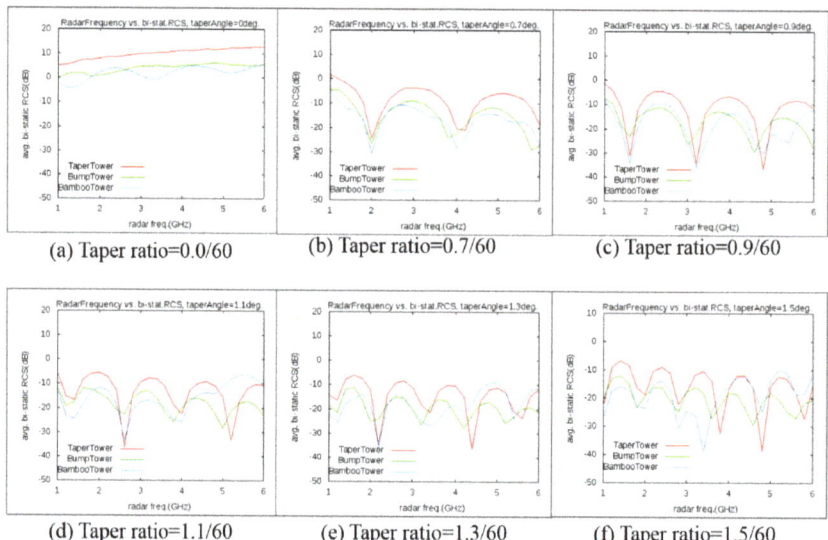

Figure 6. RCS of taper, taper bump, and taper bamboo towers under different radar frequencies and taper ratios. Red, taper towers; green, taper bump towers; blue, taper bamboo towers.

As the results show, taper bamboo towers outperform taper towers and taper bump towers if the taper ratio is less than or equal to 0.7/60. However, as the pater ratio exceeds 1.1/60, the effectiveness of taper bamboo towers deteriorates, especially for higher radar frequencies. In these cases, taper bump towers are the best reshaping method for the RCS reduction process. This result verifies the analysis presented in Section 3.2.2. As the radar frequency increases, the radar waves resemble planar light rays. The inter-ring reflections enlarge the back scatterings and damage the performance of the taper bamboo towers.

3.4. Radar Frequency and RCS

The results presented in Section 3.3 show that taper bamboo towers become less effective on alleviating average bi-static RCS as the radar frequency increases. To further study the influence of radar frequency upon the proposed reshaping methods, we performed another set of tests. In the simulations, the initial radar frequency was 1.0 GHz. Then, the radar frequency was gradually increased until it reached 10.0 GHz. The incremental value was 0.25 GHz. The taper ratios were set to 0.4/60, 0.7/60, 1.4/60, and 1.8/60, respectively. The number of bumps on the tower surfaces was decreased to six. Thus, the gaps between bumps were widened to prevent the creation of cavities. However, the bump height was increased to 10 cm to enhance scatterings. There were four towers used in this study, including a cylinder tower, a taper tower, a taper bump tower, and a taper bamboo tower.

The tests results are plotted in Figure 7. The average bi-static RCS of the cylinder, taper, taper bump, and taper bamboo tower are drawn in purple, green, blue, and yellow color, respectively. For lower taper ratios, 0.4/60 and 0.7/60, the taper bamboo tower is the most effective tower for reducing back scatterings. Nonetheless, as the taper ratio increases, its performance decreases. It generates peak average bi-static RCS at some radar frequencies. For taper ratio equals to 1.0/60, 1.4/60, and 1.8/60, the peak RCS occurs at the radar frequencies of 9.0 GHz, 6.5 GHz, and 5.0 GHz, respectively. We conjecture that the sudden increasing of average bi-static RCS is caused by the coupling effects of radar wavelength, taper ratio, and bump height. In this set of tests, the bump height was 10 cm.

The radar frequency, causing peak average bi-static RCS, decreased as the taper ratio increases. A similar phenomenon can be found in the results of Figure 6d–f, though the bump height is shorter, 8 cm, in these simulations. The magnitude of the peak average RCS is smaller.

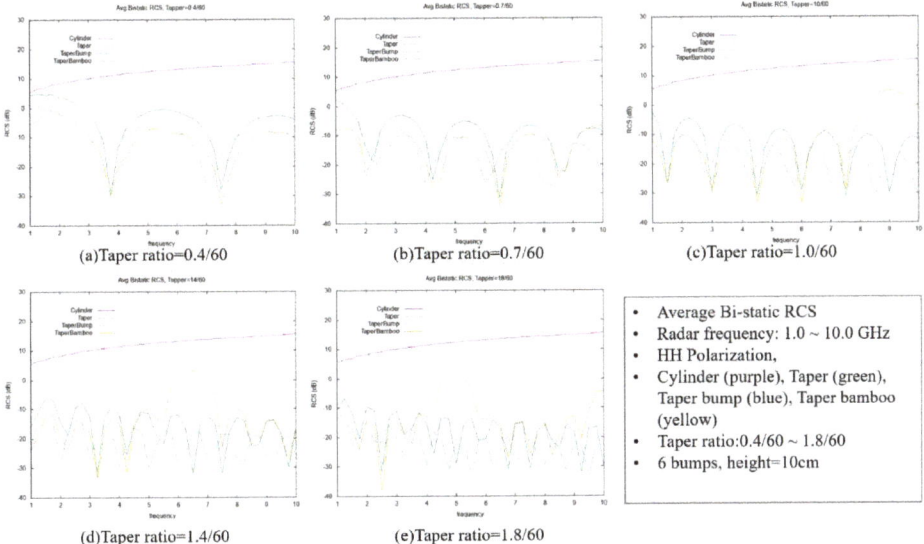

Figure 7. Average RCS of cylinder, taper, taper bump, and taper bamboo towers illuminated by radar with 1.0~10.0 frequencies.

The average bi-static RCS of the taper and taper bump tower fluctuate with radar frequency. The fluctuation becomes more frequently as the taper ratio grows. Globally, their efficiencies improve as the taper ratio and radar frequency increase. These two towers offer stable RCS reductions for all the radar frequencies and taper ratios. It is hard to compare their performances because of the fluctuation. However, in general, the taper bump tower is superior to the taper tower.

3.5. Polarization and RCS

In the previous computations, the radar waves were horizontally polarized. In order to explore the influence of polarization, we computed the RCS of towers by using not only HH but also VV polarization in another set of the experiments. The testing models included cylinder, taper, taper bump, and taper bamboo towers. The geometric parameters were fixed as follows: the taper ratio was 1.0/60; the number of bumps was six; and the bump height was 10 cm. Nonetheless, the radar frequency was set to 3, 6, and 9 GHz to study the combined effect of radar frequency and polarization. The directions and locations of the radar and the receivers were the same as the settings in Section 3.1.

The performances of the towers were evaluated by using their bi-static RCS. The bi-static RCS of the HH and VV polarization are shown in Figures 8 and 9. The bi-static RCS of the cylinder, taper, taper bump, and taper bamboo towers are rendered in purple, green, blue, and yellow color, respectively. These figures show that the bi-static RCS under HH and VV polarization are similar, especially for the cylinder, taper, and taper bamboo tower. Only the bi-static RCS of the taper bump tower are affected by the polarization method. To uncover the difference, we draw the HH and VV bi-static RCS of the taper bump tower in Figure 10. The data obtained by using HH- and VV-polarization are shaded in purple and green curves. The RCS of HH-polarization is better than that of the VV-polarization for radar frequencies equal to 3 and 9 GHz. However, the difference is not significant within the major back-scattering directions, $-20° \leq \phi \leq 20°$.

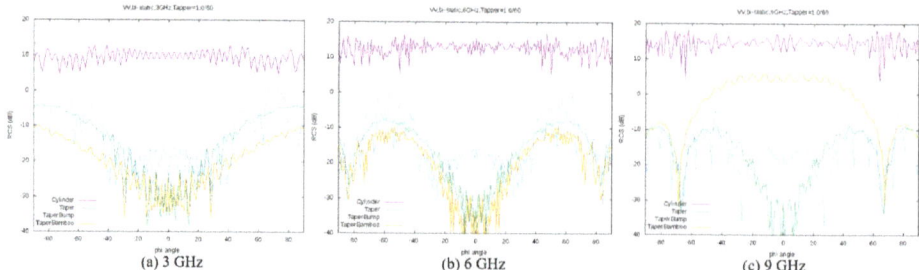

Figure 8. Bi-static RCS of the cylinder, taper, taper bump, and taper bamboo towers, illuminated by VV-polarized radar waves.

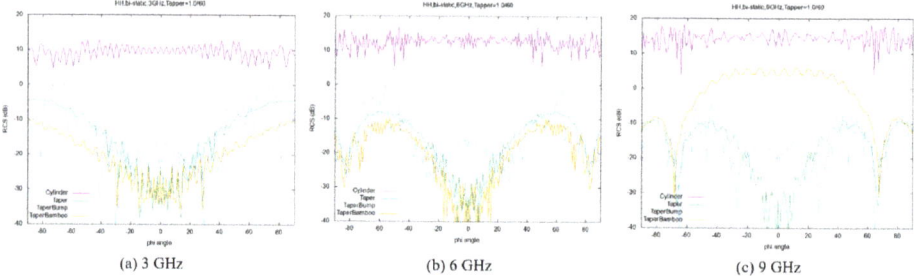

Figure 9. Bi-static RCS of the cylinder, taper, taper bump, and taper bamboo tower, illuminated by HH-polarized radar waves.

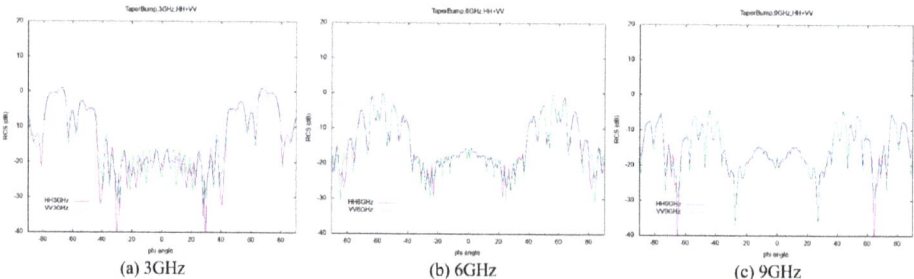

Figure 10. Bi-static RCS of the taper bump tower illuminated by HH- and VV-polarized radar waves. The purple and green curve represent HH- and VV-polarization bi-static RCS.

3.6. Discussion

In the experiments, the heights of the targets were relatively short compared with real wind turbine towers. The RCS values cannot be used to represent the radar returns of wind turbine towers. Since the main themes of this research focus on developing reshaping methods and studying the effectiveness of the proposed RCS reduction strategies, computing the RCS of a wind turbine is out of the scope of this research. However, assuming a tower is divided into N sections, the RCS values of the tower can be computed as follows:

$$B_\theta = \sum_{i=1}^{N} B_i, \quad (14)$$

where B_i is the RCS of the i-th section.

In the experiments, the incident radar waves travelled horizontally at $\theta = 90$. Their direction was orthogonal to the z-axis and parallel with the xy-plane. In reality, radar waves may hit the wind turbines at smaller zenith angles if the radar is located at higher altitudes and shorter distances. In our reshaping methods, the top end of each individual tower was modelled as a flat surface. If we decrease the zenith angle, some of the rays will hit the top end and produce radar returns. This reflectivity will be included in the resultant RCS values and obscure the contributions produced by the tapering effect and bumps. In order to prevent this problem, the zenith angle was fixed at θ = 90 in this study.

The bumps on the surface of a tower cause extra drag forces and increase the load of the tower. However, these structures enhance the mechanical strength of the tower (similar effects can be found on bamboos and tall man-made buildings). Towers can bear the added load and drag forces. In the reshaping methods, the bumps resemble waves without crests. We can generate a bumpy tower by using a flat metal sheet. At first, we press the sheet to produce vertical or horizontal bumps on the sheet surface. Then, we roll the sheet to form a bumpy tower.

4. Conclusions

Bumps and tapering effects are efficient strategies for reducing the RCS of wind turbine towers. When separately applied on the towers, each of these approaches is capable of decreasing radar returns. In this research, we combined these two methods and created hybrid wind turbine towers. These hybrid towers out-perform towers with only tapering effects or only bumpy surfaces. Test results prove that these two RCS mitigation strategies are not in mutual conflict. Instead, being put together, they produce additive effects on RCS reduction.

The taper ratio of a tower significantly influences its RCS. The higher the taper ratio, the lower the RCS. However, for a pragmatic tower, the taper ratio should be within a limit. Otherwise, the costs to build the tower would be too high. Taper bump towers possess effective RCS reduction capabilities for various radar frequencies and taper ratios. The performance of taper bamboo towers may deteriorate if the radar frequency exceeds four GHz and the taper ratio is larger than 1.0/60. However, in other cases, it is the best tower shape to reduce RCS.

The efficiencies of the taper tower and taper bamboo tower are not affected by radar polarization. On the other hand, the taper bump tower produces different bi-static RCS if we alter the polarization method. When the radar waves are horizontally polarized, its bi-static RCS are lower. However, this phenomenon is not obvious in the major back scattering directions.

Funding: This research was partially supported by the Ministry of Science and Technology, Taiwan.

Conflicts of Interest: The author declares no conflict of interest.

References

1. Matthews, J.C.G.; Sarno, C.; Herring, R. Interaction between radar systems and wind farms. In Proceedings of the Loughborough Antennas and Propagation Conference, Loughborough, UK, 17–18 March 2008; pp. 461–464.
2. Pinto, J.; Mathews, C.G.; Sarno, C. Radar signature reduction of wind turbines through the application of stealth technology. In Proceedings of the EuCAP Conference, Berlin, Germany, 23–27 March 2009; pp. 3886–3890.
3. Rashid, L.S.; Brown, A.K. Partial Treatment of Wind Turbine Blades with Radar Absorbing Materials (RAM) for RCS Reduction. In Proceedings of the Fourth EuCAP Conference, Barcelona, Spain, 12–16 April 2010; pp. 100–110.
4. Rashid, L.S.; Brown, A.K. Radar cross-section analysis of wind turbine blades with radar absorbing materials. In Proceedings of the European Radar Conference (EuRAD), Manchester, UK, 12–14 October 2011; pp. 97–100.
5. Ling, H.; Chou, R.-C.; Lee, S.-W. Shooting and Bouncing Rays: Calculating the RCS of an arbitrarily shaped cavity. *IEEE Trans. Antennas Propag.* **1989**, *37*, 194–205. [CrossRef]
6. Ueng, S.K.; Chan, Y.H. A Reshaping method for wind turbine RCS reduction. *Appl. Mech. Mater.* **2015**, *764*, 457–461. [CrossRef]
7. Ueng, S.K.; Chan, Y.H.; Lu, W.H.; Chang, H.W. Radar return reduction for wind turbines by using bump structures. *Sains Malays.* **2015**, *44*, 1701–1706.

8. Ueng, S.K.; Chen, Y.W. Scattering Reduction Using Bumps and Tapering Effect. In Proceedings of the ICCPE2016 Conference, Kenting, Taiwan, 30 September–3 October 2016.
9. Hou, Y.C.; Liao, W.J.; Ke, J.F.; Zhang, Z.C. Broadband and broad-angle dielectric-loaded RCS reduction structures. *IEEE Trans. Antennas Propag.* **2019**, *67*, 3334–3345. [CrossRef]
10. Song, J.; Wu, X.; Huang, C.; Yang, J.; Ji, C.; Zhang, C.; Luo, X. Broadband and tunable RCS reduction using high-order reflections and salisbury-type absorption mechanisms. *Sci. Rep.* **2019**, *9*, 9036. [CrossRef] [PubMed]
11. Chen, J.; Cheng, Q.; Zhao, J.; Dong, D.S.; Cui, T.J. Reduction of radar cross section based on a metasurface. *Prog. Electromagn. Res.* **2014**, *146*, 71–76. [CrossRef]
12. Mighani, M.; Dadashzadeh, G. Broadband RCS reduction using a novel double layer chessboard AMC surface. *Electron. Lett.* **2016**, *52*, 1253–1255. [CrossRef]
13. Baldauf, J.; Lee, S.W.; Lin, L.; Jeng, S.K. High frequency scattering from trihedral corner reflectors and other benchmark targets: SBR versus experiment. *IEEE Trans. Antennas Propag.* **1991**, *39*, 1345–1351. [CrossRef]

© 2020 by the author. Licensee MDPI, Basel, Switzerland. This article is an open access article distributed under the terms and conditions of the Creative Commons Attribution (CC BY) license (http://creativecommons.org/licenses/by/4.0/).

Article

Electronically Controlled Actuators for a Micro Wind Turbine Furling Mechanism †

Mihai Chirca [1], Marius Dranca [1], Claudiu Alexandru Oprea [1], Petre-Dorel Teodosescu [1], Alexandru Madalin Pacuraru [1], Calin Neamtu [2] and Stefan Breban [1,*]

[1] Department of Electric Machines and Drives, Technical University of Cluj-Napoca, 400114 Cluj-Napoca, Romania; mihai.chirca@mae.utcluj.ro (M.C.); marius.dranca@mae.utcluj.ro (M.D.); Claudiu.Oprea@emd.utcluj.ro (C.A.O.); petre.teodosescu@emd.utcluj.ro (P.-D.T.); Alexandru.Pacuraru@mae.utcluj.ro (A.M.P.)
[2] Department of Design Engineering and Robotics, Technical University of Cluj-Napoca, 400114 Cluj-Napoca, Romania; calin.neamtu@muri.utcluj.ro
* Correspondence: Stefan.Breban@emd.utcluj.ro
† This paper is an extended version of our paper published in 2019 at 11th International Symposium on Advanced Topics in Electrical Engineering (ATEE), Bucharest, Romania, 28–30 March 2019, doi:10.1109/ATEE.2019.8724908.

Received: 30 June 2020; Accepted: 12 August 2020; Published: 14 August 2020

Abstract: This paper presents two electromechanical systems used for the overspeed protection of small wind turbines. The actuators have the purpose of rotating the back rudder (tail vane) of the wind turbine when the blades are overspeeding. The rudder rotation angle is 90 degrees in order to completely turn the wind turbine blades away from the wind flow direction. The first device is a new limited-angle torque electromechanical actuator consisting of a device with a simplified structure composed of four permanent magnets (two on each side) glued on a rotor mounted between two stator poles built from ordinary rectangular construction pipes and an electronic control unit. The second device is based on a regular stepper motor actuator with a reduction gear and an appropriate control scheme to maximize the energy harvested at high, over-nominal wind speeds. A generic comparison is provided for the proposed solutions.

Keywords: furling; limited-angle torque actuator; stepper motor; micro wind turbine; finite element analysis

1. Introduction

Small wind turbines are struggling to occupy a share of the renewable energy market for residential areas. This is due to the continuous reduction in photovoltaics prices and relatively low average wind speed in most urban areas. Thus, the installation of small and micro wind turbines is limited to locations where the average wind speed is above 5–7 m/s. In order to change this situation, new topologies of urban wind turbines should be developed with the following characteristics: low cost per kWh produced, large surface to convert even low wind speeds into useful energy, low noise operation, and harmless to birds and bats. Usually, wind-direction-orientable turbines have higher efficiencies compared with turbines working in winds from any direction [1] (Figure 1). Furthermore, with a back rudder, the wind turbines can be turned out of the wind. Thus, developing effective braking mechanisms allows for the safe operation of small wind turbines. In order to avoid possible damage to the wind turbine, or even to nearby peoples or animals, a cheap and safe system that allows for the reduction of the rotor speed is required. The most common solutions can be grouped into three main categories: mechanical, aerodynamic, and electric brakes.

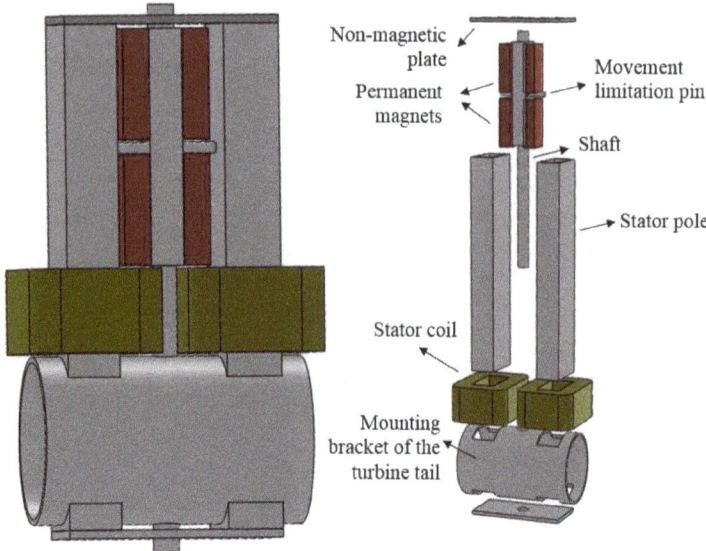

Figure 1. Structure of the limited-angle torque (LAT) actuator.

Widely used in wind turbines, mechanical brakes suffer from frequent wear, which requires a high maintenance cost or even component replacements. Disk brakes are generally used to stop the wind turbine rotor via friction. Despite this, mechanical braking systems are required in large wind turbines during maintenance or as back-up braking in the case of extreme conditions.

Another solution is controlling the pitch angle of the blades at high wind speeds and thus increasing the angle of attack so the lift force is reduced, a method known as aerodynamic braking. Even though this system provides more control options for the system, it adds weight to the nacelle. Aerodynamic brakes imply high costs but can be used a very high number of times, making them a good solution for high-power systems.

A comparison between soft-stall control and furling control is presented in Muljiadi et al. [2]. In terms of energy production during high winds, soft-stall control is more advantageous.

The rotational speed of the wind turbine can be reduced by controlling the level of interaction between the magnetic fields in the stator and rotor of the generator, which are usually produced by windings and permanent magnets, respectively. The amount of braking force can be adjusted by inserting a variable resistor in the stator electric circuit and is proportional to the electric current passing the windings. More advanced control strategies can be used to adjust the generator load using brake resistors (or thermistors) and power electronics components. A redundant electrical braking system for wind turbine generators was proposed by Wang et al. [3], which consists of a hybrid system that uses a dynamic brake resistor connected on the system's Direct-Current (DC) link and a dump load resistor connected to the generator terminals, which provides several advantages compared to conventional braking methods, such as higher reliability, modularity, and a higher degree of control. Matsui [4] and Sugawara [5] introduced a cheaper braking method that uses a "Y" connection circuit with negative temperature coefficient (NTC) thermistors that are directly connected to the generator terminals. While in conventional electric braking systems, the generator stator windings are short-circuited to abruptly increase the current, this solution allows for a smoother reduction of the rotating speed of the wind turbine. Gong et al. [6] proposed another safe and cheap method for controlling the braking of a micro wind turbine by using a Pulsed-Width Modulation (PWM)-controlled braking circuit with field-effect transistors (FETs). These solutions are simple and effective but have a drawback: due to high currents in the generator windings, the inside temperature increases and the stator magnetic field is strong. This can lead to partial demagnetization of the permanent magnets if the "knee point" of the demagnetization characteristic is attained. This happens especially if the permanent magnets are mounted on the surface of the rotor.

Several solutions for rotary actuators were proposed with different shapes and operating principles and using gear systems to transmit the rotational motion, such as the one developed by Cuches [7]. Different rotor designs were proposed in Cuches et al. [8] that use mechanical springs to provide torque at a certain desired position in the absence of electrical current through the motor. Different types of electromechanical actuators are presented in Oudet and Prudham [9], showing a torque variation that is approximately independent of the rotor's angular position. An actuator with two stable stationary positions when no current is fed to the windings is introduced in Biwersi and Gandel [10]; the major disadvantage of this solution, which employs only one winding, is that a supply voltage with both polarities is required to move the rotor from one stable position to the other. A unipolar power source can be used if two or more windings are considered and two differential control signals are required to determine the motion of the rotor between the two no-current stable positions.

Regarding the yaw displacement or furling, a solution is detailed in Mohammadi et al. [11], where complex angular movement control is proposed in order to protect the wind turbine and maximize the energy harvested at high winds exceeding the nominal values. In References [12,13], a limited-angle torque (LAT) actuator with a similar construction to the one shown in Figure 1 was presented.

The paper is structured as follows: Section 2 presents the design of the LAT electromechanical actuator and the electronic circuit used to control the LAT actuator, Section 3 provides details regarding the stepper motor actuator and the electronic circuit for angular control, and in Section 4, a discussion regarding the solutions is presented.

2. Limited-Angle Torque Electromechanical Actuator

2.1. Design of the LAT Electromechanical Actuator

The LAT electromechanical actuator was designed to ease the manufacturing process so it can be made by small workshops, reducing the production cost. The developed prototype, presented in Figure 1, was used for a 2.5 kW, 500 rpm wind turbine. The main purpose of this device is to angle the wind turbine blades from facing a direct wind flow if the electric generator is surpassing a certain speed and reaches a threshold voltage, or if a loss of electrical load emerges. This device has a simple and robust design that allows for the furling of the wind turbine. It consists of a rotor having four radially magnetized permanent magnets, mounted on a shaft by gluing. Two copper windings are placed around two stator poles and are made from ordinary rectangular steel profiles. The stator poles are fixed to the tail mounting support (Figure 2), which is also made of steel, allowing for closing the magnetic circuit. A non-magnetic material plate (a 3D printed element was used for the prototype) fixed to the ends of the rectangular steel profiles supports the bearing, together forming a rigid assembly. The rotor movement is limited by a pin mounted through the shaft.

The supply voltage required for the electronic circuit, which powers the winding of the electromechanical actuator, is directly extracted from the generator three-phase rectifier assembly. If the wind generator speed exceeds the limit (voltage threshold is reached), the moving part will rotate from one stationary position to the other, thus furling the wind turbine.

As shown in Figure 2, the LAT actuator is mounted on a wind turbine tail using a dedicated mounting support. During normal operation, when the wind speed value is below the threshold, the nacelle and the tail are aligned with the direction of the wind. For higher wind speed values, once the generated voltage exceeds a certain limit, the LAT is activated, causing the tail to rotate and to form a 90° angle with the rest of the wind turbine. In this case, the nacelle is no longer aligned with the wind direction and the blades are not driven efficiently by the wind, causing a decrease in the rotating speed of the generator.

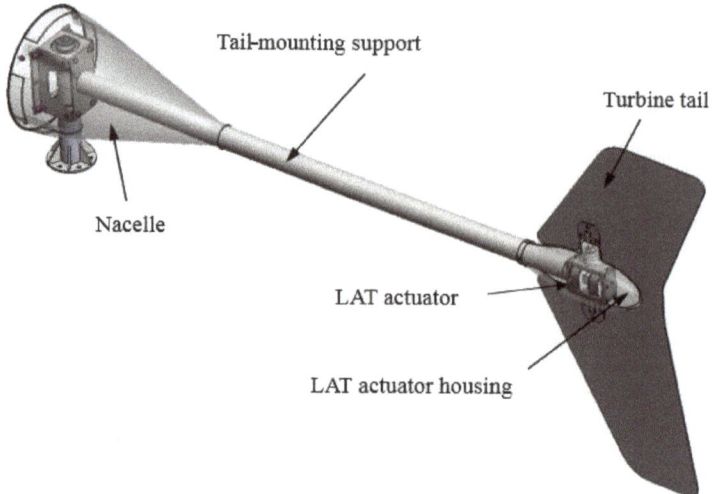

Figure 2. LAT actuator placement in the wind turbine assembly.

To investigate the performances of the limited-angle torque electromechanical actuator, design analysis was conducted by means of a 3D finite element method (FEM) model (Figure 3) using the commercial software JMAG-Designer, version 19, JSOL Corporation, Tokyo, Japan. Previous studies [13,14] have shown that 3D FEM analysis provides more precise results compared to the simplified 2D FEM analysis, especially for complex structures, such as the proposed LAT.

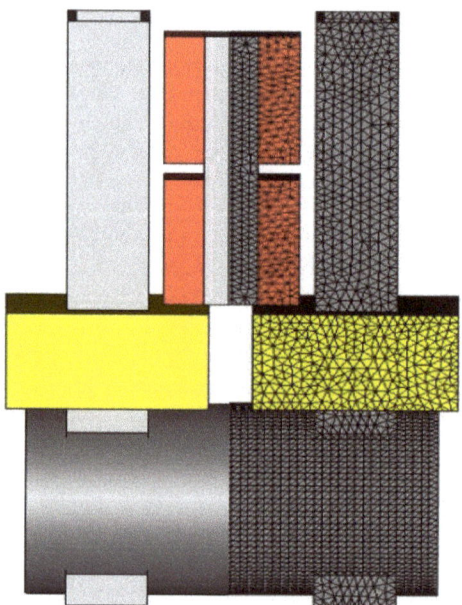

Figure 3. LAT 3D model and mesh in JMAG Designer.

Figure 4 is used to present the operating mode for the limited 90° rotation of the electromechanical actuator: in the initial position of the actuator rotor, presented in Figure 4a, the magnetization axis of the four permanent magnets forms a 45° angle to the magnetization axis of the stator poles.

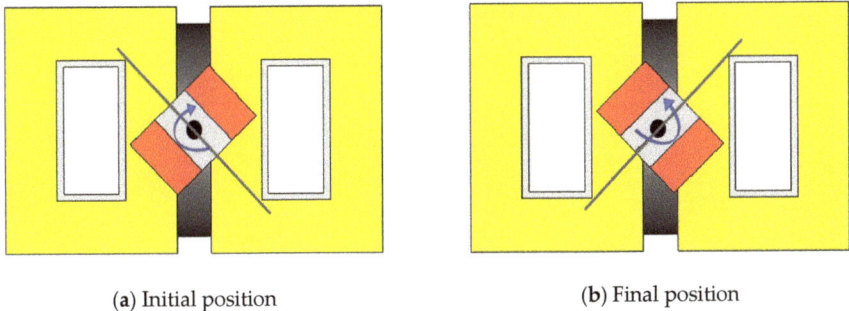

(a) Initial position   (b) Final position

Figure 4. Upper view of the electromechanical LAT actuator (**a**) initial and (**b**) final rotor positions.

In this position, when no current is fed to the windings, a negative cogging torque is obtained, as shown in Figure 5, which tries to align the magnetization axis of the rotor and the stator. The cogging torque component has a value of approximately −3 Nm, at initial position, which is enough to keep the tail vane aligned with the nacelle and the wind direction.

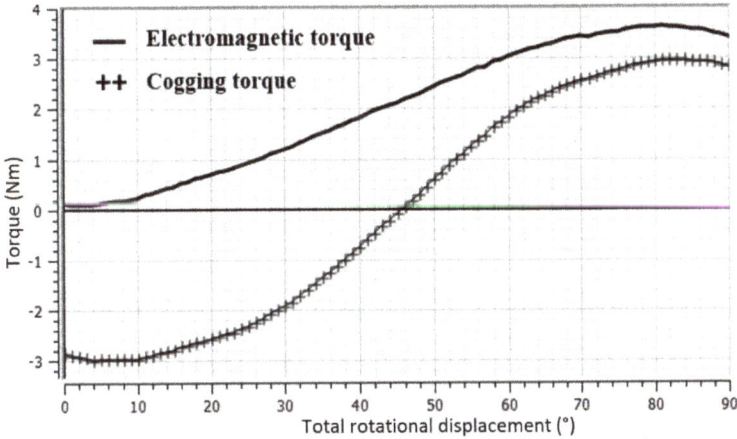

Figure 5. Electromagnetic and cogging torque obtained from the 3D analysis.

In the case where the voltage threshold is exceeded, the windings are fed with a DC voltage and the LAT actuator develops a positive electromagnetic torque. Thus, the rotor moves from the initial (Figure 4a) position to the final position shown in Figure 4b, rotating from 0° to 90°; this position is maintained even if the winding is not fed due to positive electromagnetic and cogging torque.

To return to the initial position (Figure 4a), corresponding with normal wind turbine operation, the windings are fed with a reverse polarity voltage from the continuous supply voltage, and the actuator rotor returns from the final position (Figure 4b), backward throughout the rotation range (from 90° to 0°).

The electrical circuit consists of two series-connected coils that are fed from a continuous voltage source; the main properties are included in Table 1. Each coil has 450 turns and a coil resistance of 2.8 Ω. While the coils are fed, during the furling process initiated by the movement of the rotor, a 13.34 A current (Root Mean Square (RMS) value) was obtained in the 3D FEM analysis. High values of Joule losses were recorded in the stator coils of about 1000 W. However, there is no major concern about coil insulation damage due to the short time of operation (under 1 second).

Table 1. Electrical circuit properties.

Parameter	Value
Supply voltage	200 V
Number of turns/coil	450
Coil resistance/coil	2.8 Ω
Rated current RMS	13.34 A
Coil connection type	Series

Figure 6 presents the magnetic flux density distribution obtained by means of 3D FEM analysis, showing that there were small saturated regions in the stator poles with values over 2.2 T. Only one of the two coils is shown to allow for the visualization of the magnetic flux density under the coils.

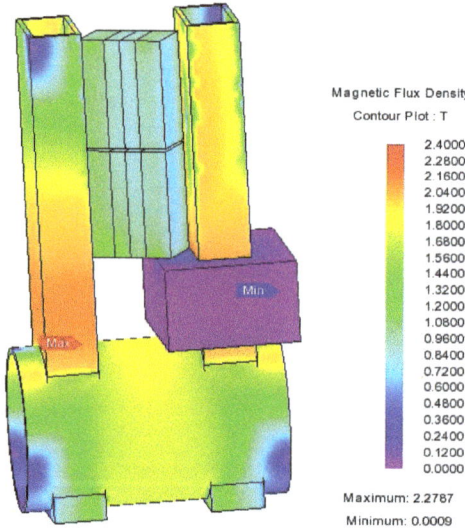

Figure 6. 3D magnetic flux density distribution in the LAT.

The electronic control device presented in the following section only consumes electrical power while the LAT is in the second stationary position and when the device returns from this position to the initial position. Due to the negative torque given by the cogging component, the system is stable when no current is fed to the stator coils.

2.2. Electronic Circuit for LAT Electromechanical Actuator Control

Based on the principle of an ON-OFF command, with the help of the electromechanical actuator developed and presented in the previous section, Figure 7 presents the practical implementation of a bipolar converter under different working stages and LED signalizations. Figure 8 shows the electronic schematics of the proposed solution that is based on the "comparator stage," "detection stage," "fixed-pulse stage," "shoot-through protection stage," "power inverter stage," and the LAT electromechanical actuator. The initial and final positions are considered to be 0° and 90°, respectively, for the control strategy explained bellow.

(a) (b)

Figure 7. Bipolar converter images under different detection stages: (**a**) minimum voltage level reached and the 0° movement is performed, and (**b**) the maximum voltage level reached and the 90° movement is performed.

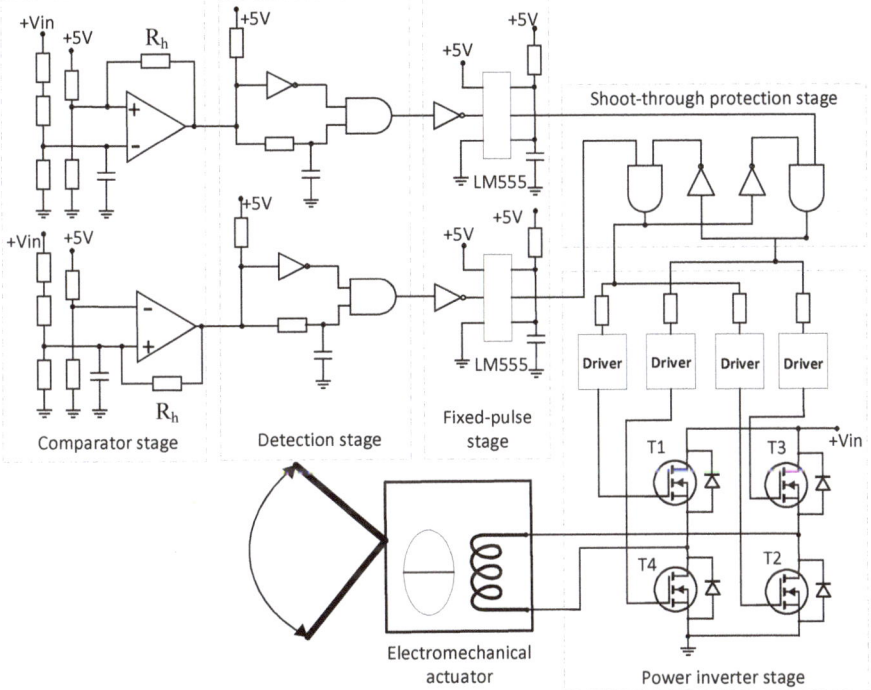

Figure 8. Current flow schematic for the proposed angular bipolar control.

The comparator stage is based on the monitoring of the input voltage and evaluation with two predefined values:

- 190 V—minimum detection voltage (V_{min})
- 210 V—maximum detection voltage (V_{max}).

By using two comparators, the system limits the LAT oscillation at high speeds because between these voltage levels, the actuator is not performing any actions. These detection levels can be changed in order to cope with specific applications and dynamics.

The "detection stage" and the "fixed-pulse stage" are used to determine the right activation time for the command signals and the duration of the pulse. For protection purposes, the "shoot-through protection stage" is implemented to avoid triggering the power stage transistors (T1–T4 or T2–T3) at the same time when very fast input voltage changes are detected. The power inverter stage is developed on a regular H-bridge inverter that performs bipolar voltage across the electromechanical actuator coil.

For the initial testing presented in Figure 9, the "detection stage," the "fixed-pulse stage," and the feedback comparator resistances R_h were not used. Thus, in Figure 9a, the input voltage reached the maximum voltage detection level V_{max} and a command pulse was obtained that triggered the T1 and T2 transistors, applying a positive rectangular voltage across the actuator inductor, as can be seen in Figure 9b. Similarly, in Figure 9c,d, the minimum voltage detection level V_{min} caused a command signal to be applied to the T3 and T4 transistors and a negative rectangular voltage to be applied across the actuator inductor. In Figure 9e, a detection sequence of the minimum and maximum voltages was being considered, obtaining the command signals, while in Figure 9f, the bipolar voltage across the inductor can be observed. It is important to specify that all the results presented in Figure 9 were individually obtained for different practical experiments with diverse time actions of the input voltage, thus the widths of the command pulse and actuator voltages do not necessarily match between the presented oscilloscope pictures.

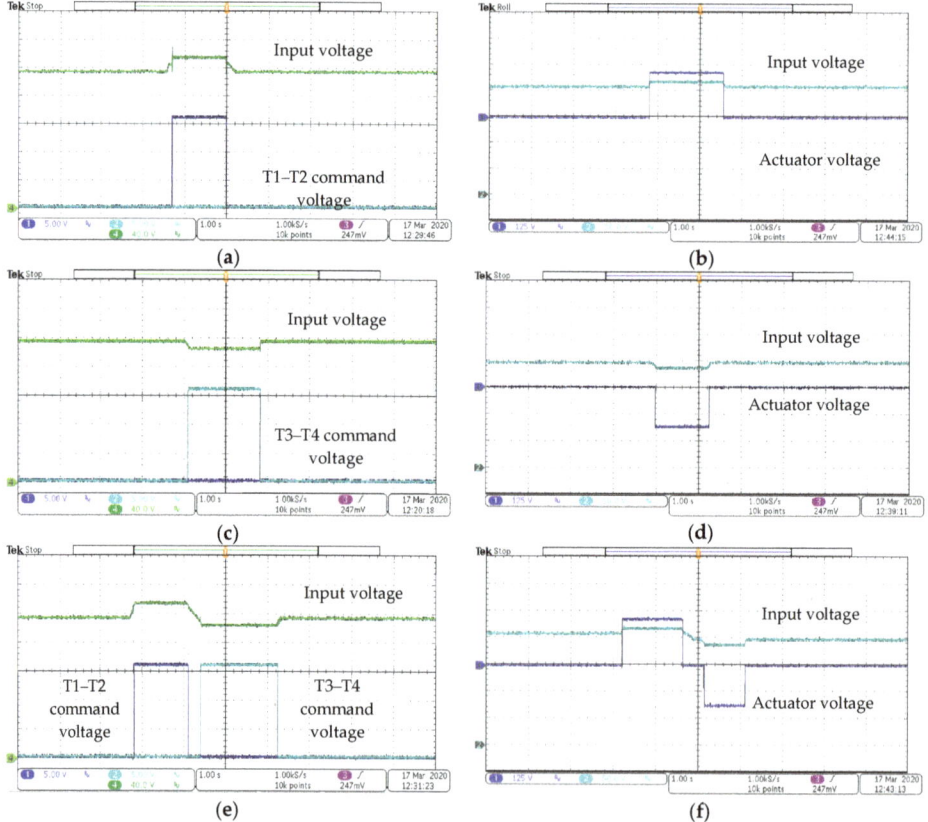

Figure 9. Working principle waveforms of the bipolar current angular command: (**a**) overvoltage transistor command signal, (**b**) overvoltage coil voltage, (**c**) undervoltage transistor command signal, (**d**) undervoltage coil voltage, (**e**) overvoltage and undervoltage transistor commands, and (**f**) overvoltage and undervoltage bipolar coil voltage.

Under slow variations of the input voltage, the comparator stage was performing incorrectly, as can be seen in Figure 10a, where multiple undesired switchings were noticed. One solution was to perform local feedback using the high resistance R_h from Figure 8. The resistance R_h implementation results can be observed in Figure 10b.

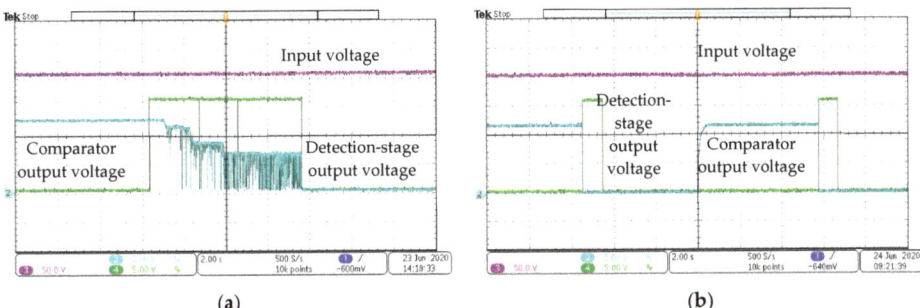

Figure 10. Working principle waveforms of the bipolar current angular transistor command under slow slope changes of the input voltage: (**a**) no stabilization method (no R_h resistance) and (**b**) with a feedback stabilization method (with R_h resistance).

As one can notice from Figure 9, the transistors' command signal durations were always correlated with the duration of the overvoltage or undervoltage, which is a negative aspect because the coil can be thermally damaged if the current is applied for a long period. To cope with this problem, the duration of the command pulse needed to be limited, thus the usage of the "detection stage" and the "fixed-pulse stage" from Figure 8 can be one way of increasing the performance of the system. As can be seen in Figure 10, the "detection stage" was used to perform a pulse at the output that determined the crossing detection of the minimum and maximum values. The "fixed-pulse stage" defines the duration of this pulse so that the coil will be powered long enough to perform the coil switching from 0° to 90° and from 90° to 0° without thermally overloading the inductor windings. The results obtained by using all the improvements can be observed in Figure 11, where fixed pulses were applied for every detection. Figure 11a,b show that the time the transistor command voltages were applied was not triggered by the duration of the overvoltage/undervoltage values of the input voltage. Moreover, the command voltage of the T1–T2 transistor was intentionally set to be slightly bigger than the T3–T4 command to show that this time can be controlled via the "fixed-pulse stage."

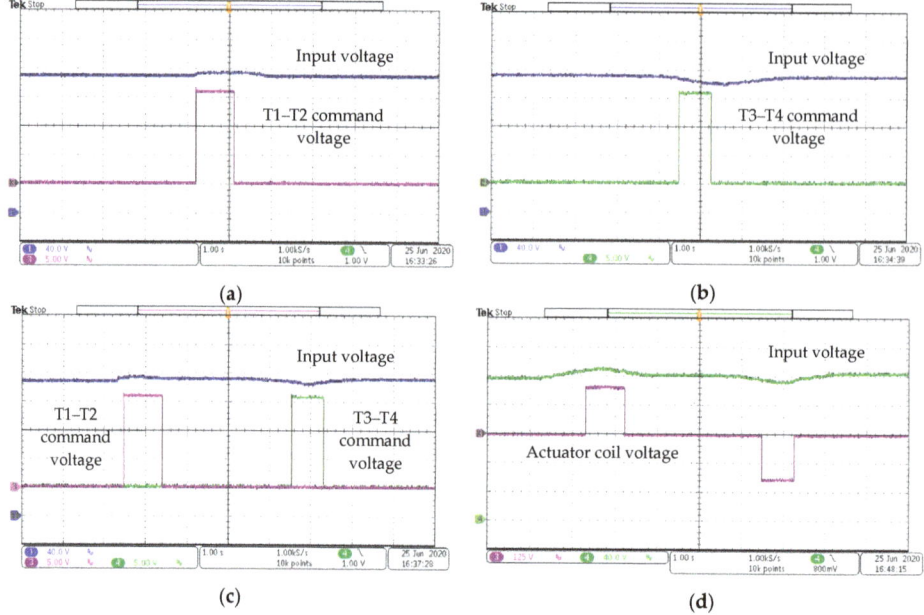

Figure 11. Working principle waveforms of the bipolar current angular command with a limited activation time: (**a**) overvoltage transistor command signal, (**b**) undervoltage transistor command signal, (**c**) overvoltage and undervoltage transistor commands, and (**d**) overvoltage and undervoltage bipolar coil voltage.

3. Geared Stepper Motor Actuator and the Electronic Circuit for Angular Control

Based on the principle of angular control, with the help of a stepper motor, Figure 12 shows two electronic circuits that were developed for the proper management of the detection cases and one maximum 90° angular element driven by the motor. The circuit from Figure 12a was based on a comparator stage that monitored the input voltage and associated the value with four predefined levels, as can be seen in Figure 13 in the "comparator stage." Based on the proper selection of the comparator resistances, the predefined voltage levels were:

- 190 V—minimum detection voltage (V_{min})
- 200 V—nominal voltage for the detection stage (V_{nom})
- 210 V—first maximum voltage level ($V_{max.1}$)
- 215 V—second maximum voltage level (for supplementary protection) ($V_{max.2}$).

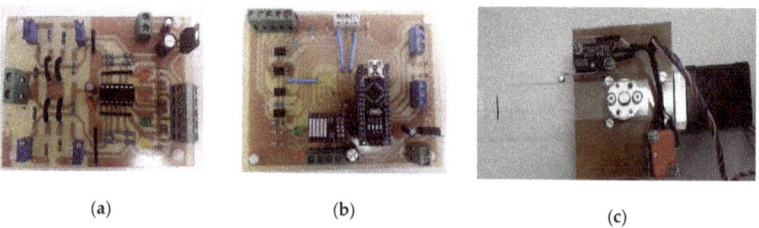

Figure 12. Developed elements for the stepper motor angular application: (**a**) input voltage level detection stage, (**b**) digital control and power stage, and (**c**) 90° stepper motor driven movement element.

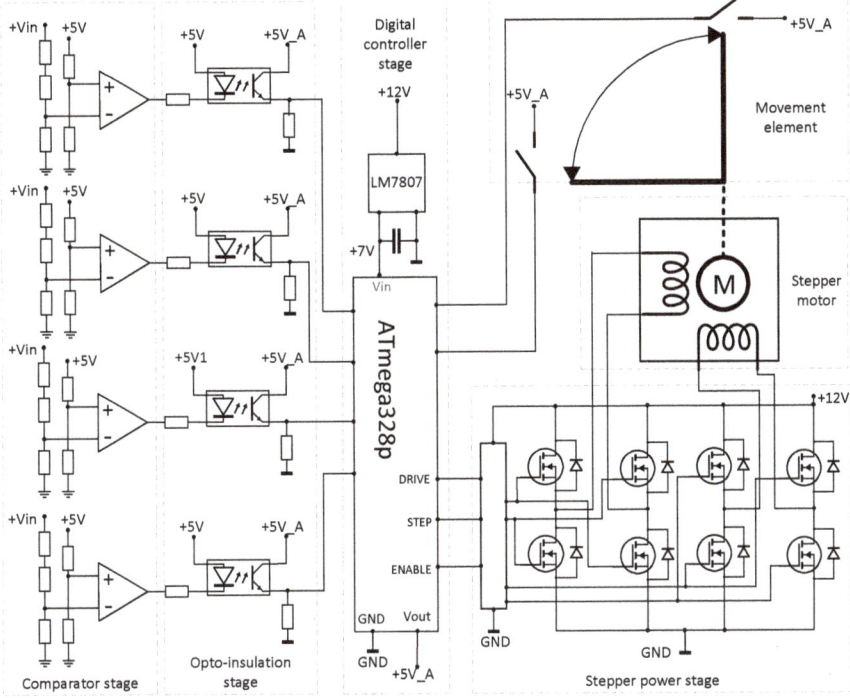

Figure 13. Current flow schematic for the proposed angular motor control.

The electronic circuit from Figure 12b was developed by using the "opto-insulation stage," "digital controller stage," and the "stepper power stage" presented in Figure 13. Using the opto-isolated signals from the comparators, the digital controller stage performed certain analyses in order to develop the right command signals for the stepper motor driver. The motor driver was controlled with three signals in direct correlation with the overvoltage/undervoltage detection stages delimitated in Figure 14e:

- Enable: A signal that activates both H-bridge circuits when the input state is low, and when it is high, the H-bridges are locked; the enable signal is in a high state until the overvoltage detection stage starts, meaning that the $V_{max.1}$ value of the input voltage is reached. During the overvoltage detection stage, the output signal is low until the V_{nom} level is reached. Similarly, the high state is present until the undervoltage detection stage starts, meaning that the V_{min} value of the input voltage is reached. During the undervoltage detection stage, the output signal is low until the V_{nom} level is reached again.
- Drive: A signal by which the rotation direction of the stepper motor can be changed. If the state is high, then the motor rotates clockwise. This is obtained during the overvoltage detection stage and when the $V_{max.1}$ voltage level is reached. If the $V_{max.2}$ voltage is also reached, it means that the stepper needs to perform all the rotational steps to obtain a 90° angle as quickly as possible. If the drive signal is in the low state, the motor rotates anticlockwise. This is obtained during the undervoltage detection stage and when the V_{min} level is reached.
- Step: A signal that represents the command pulse or the increment motor step. The slope of the incremental steps is set to a low value so that it responds adequately in relation to the usual voltage behavior. For protection purposes, if the value $V_{max.2}$ is reached, the slope time is increased so that the 90° angle is reached as soon as possible.

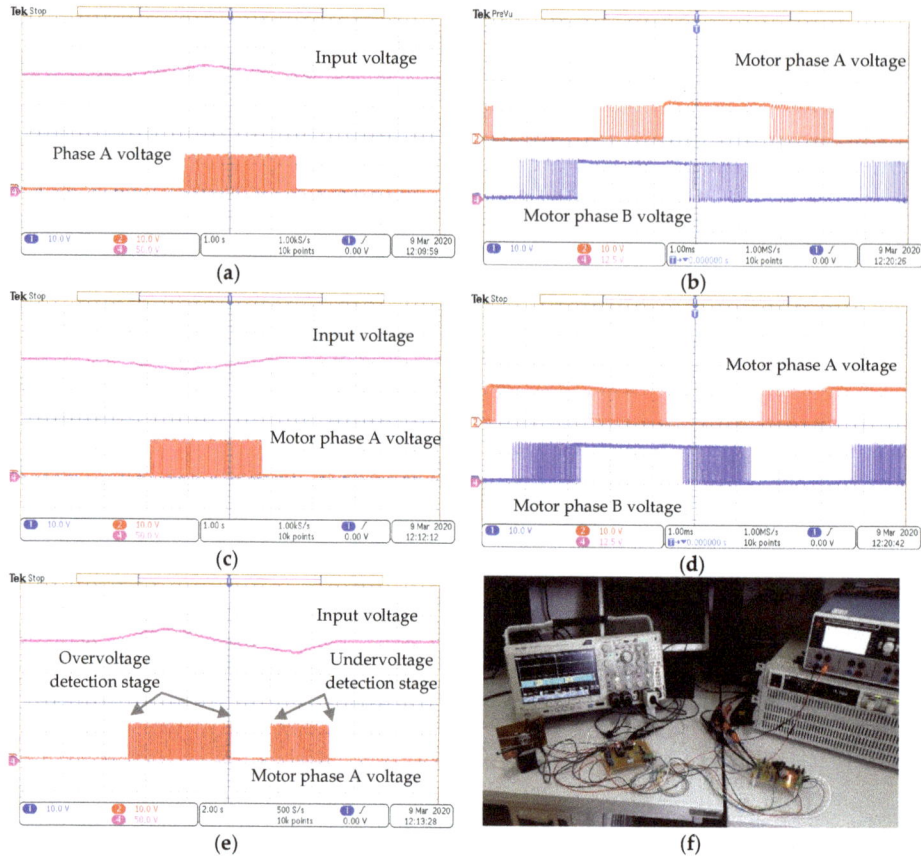

Figure 14. Working principle waveforms of the stepper motor angular command: (**a**) overvoltage command signal, (**b**) voltages of the motor phases in the overvoltage detection, (**c**) undervoltage transistor command signal, (**d**) voltages of the motor phases in the undervoltage detection, (**e**) overvoltage and undervoltage commands, and (**f**) laboratory test setup.

Figure 12c shows the stepper motor angular movement element that presents two microcontacts for the hardware angle limitation of the actuator. The contacts are also used for the software recalibration of the position element after a possible restart of the microcontroller.

In the following, the experimental results are presented, where Figure 14 shows the oscilloscope waveforms for different working stages of the actuator system. In Figure 14a, the clockwise movement was performed for a time (overvoltage detection stage) after the $V_{max.1}$ voltage was reached, and the wind system was protected against overvoltage. Figure 14b shows the motor phase voltage sequence for this working stage. In Figure 14c,d, the same information is presented, but in this case, for the anticlockwise movement (undervoltage detection stage), the wind system voltage was working at an output voltage smaller than the V_{nom} value. Figure 14e combines both clockwise and anticlockwise movements, while Figure 14f shows the test bench setup.

Outside the overvoltage/undervoltage detection stages, no angular movement was completed and the digital controller was not performing any movement command until one of the two limits (V_{min} and $V_{max.1}$) was reached. This case can be observed in Figure 14e.

4. Discussion

As can be seen in the presented results, in the case of the bipolar controlled actuator, the proposed solution displayed good dynamic performance with simple electronic schematics. For the input voltage between the minimum and maximum detection levels (V_{min} and V_{max}), the bipolar electromechanical actuator did not perform any movements; thus, to some extent, it improved the energy harvested at high wind speeds. Further improvements could be achieved in terms of simplifying the schematics of the practical layout by employing specialized integrated circuits that follow the logic schematics presented in the paper. One such example could be the use of the integrated circuit LTC6993 [15] to replace the "detection stage" and the "fixed-pulse stage." This type of actuator has two major advantages: lower cost and higher reliability. The drawback is represented by the limited wind turbine power regulation capability, which is only achievable by repeating the movement of the rotor between initial and final positions.

Regarding the stepper motor solution, the electronic system performed well and had a unique inherent capability, where under slow changes of input voltage, it acted as an active maximum power limitation. Thus, the tail angle can be changed continuously so that it keeps the wind power around the nominal one. The predefined voltage levels can be rearranged, and the incremental steps characteristics can be also changed to the developer's needs. Future work can be focused on decreasing the number of detection levels by eliminating the $V_{max.2}$ voltage level. In this case, further control steps are needed to measure and analyze the input voltage slope changes in order to impose the right motor incremental step. This can lead to a more complex control algorithm but with benefits in terms of further increasing the energy harvested at the nominal rotation speed of the wind turbine. This stepper-motor-based actuator has higher implementation costs, and due to its complexity, is more prone to failures, but can allow the wind turbine to work at the rated power, even in high winds.

Table 2 gives a relative comparison of the solution detailed in this paper.

Table 2. Comparison of the proposed solutions.

Overspeed Protection System	ON-OFF Bipolar Actuator	Geared Stepper Motor-Based Actuator
Cost	low	low
Actuator complexity	low	medium
Electronic control complexity	low	medium
Energy harvested at high speed	medium	high
Power consumption when active	high	low
Reliability	high	medium
Weight	high	low
Volume	high	low

To conclude, the two solutions presented in this paper each have advantages and drawbacks but help solve an important issue regarding the protection of small wind turbines in high wind conditions or the loss of electrical load.

5. Patents

Breban, S.; Teodosescu, P.D.; Neag, A. V.; Chirca, M. Electromechanical actuator with electronic control device 2018, RO131166B1.

Author Contributions: Conceptualization, S.B.; methodology, S.B. and P.-D.T.; Simulations, M.C., M.D. and A.M.P.; validation, M.C., M.D., C.N. and A.M.P; writing—original draft preparation, M.C., M.D., C.A.O., P.-D.T., A.M.P. and S.B.; writing—review and editing, S.B., C.A.O. and P.-D.T. All authors have read and agreed to the published version of the manuscript.

Funding: This research was funded by the European Regional Development Fund through the Competitiveness Operational Program 2014–2020, Romania, grant number 16/1.09.2016, project name: "High power density and high-efficiency micro-inverters for renewable energy sources—MICROINV."

Conflicts of Interest: The authors declare no conflict of interest.

References

1. Mohammad, E.; Rouzbeh, S.; Rezvan, A.; Mostafa, S.S. Numerical Investigation of the Savonius Vertical Axis Wind Turbine and Evaluation of the Effect of the Overlap Parameter in Both Horizontal and Vertical Directions on Its Performance. *Symmetry* **2019**, *11*, 821.
2. Muljiadi, E.; Forsyth, T.; Butterfield, C.P. Soft-Stall Control Versus Furling Control for Small Wind Turbine Power Regulation. In Proceedings of the Windpower '98, Bakersfield, CA, USA, 27 April–1 May 1998.
3. Wang, T.; Yang, W.; Yuan, X.; Teichmann, R. A Redundant Electrical Braking System for Wind Turbine Generators. In Proceedings of the European Conference on Power Electronics and Applications, Aalborg, Denmark, 2–5 September 2008.
4. Matsui, Y.; Sugawara, A.; Sato, S.; Takeda, T.; Ogura, K. Braking Circuit of Small Wind Turbine Using NTC Thermistor under Natural Wind Condition. In Proceedings of the 7th International Conference on Power Electronics and Drive Systems, Bangkok, Thailand, 27–30 November 2007.
5. Sugawara, A.; Yamamoto, K.; Yoshimi, T.; Sato, S.; Tsurumaki, A.; Ito, T. Research for Electric Brake Using NTC Thermistors on Micro Wind Turbine. In Proceedings of the 12th International Power Electronics and Motion Control Conference, Portoroz, Slovenia, 30 August–1 September 2006; pp. 1597–1601.
6. Gong, S.; Gao, C.; Chen, Z.; Zhang, J.; Ji, F. Research on Gentle Electric Brakes Using PWM and FET Control Circuits on Micro Wind Turbines. In Proceedings of the IEEE 2nd Advanced Information Technology, Electronic and Automation Control Conference (IAEAC), Chongqing, China, 25–26 March 2017; pp. 1737–1741.
7. Cuches, J.P. Rotary Actuator. U.S. Patent 3039027A, 12 June 1962.
8. Cuches, J.P.; Wiiage, M.; Henry, R. Angular Displacement Solenoid. U.S. Patent 3221191, 30 November 1965.
9. Oudet, C.; Prudham, D. Monophase Electromagnetic Rotary Actuator of Travel between 60 and 120 Degrees. U.S. Patent 5334893A, 2 August 1994.
10. Biwersi, S.; Gandel, P. Hybrid Single-Phase Bistable Rotary Actuator. E.P. Patent 1581991B1, 13 March 2013.
11. Mohammadi, E.; Fadaeinedjad, R.; Moschopoulos, G. A Study of Power Electronic Based Stall and Electromechanical Yaw Power Control Strategies in Small-Scale Grid-Connected Wind Turbines. In Proceedings of the 2018 IEEE Applied Power Electronics Conference and Exposition (APEC), San Antonio, TX, USA, 4–8 March 2018; pp. 2323–2329. [CrossRef]
12. Breban, S.; Teodosescu, P.D.; Neag, A.V.; Chirca, M. Electromechanical Actuator with Electronic Control Device 2018. Romanian Patent 131166B1, 30 August 2018.
13. Chirca, M.; Drancă, M.; Teodosescu, P.D.; Breban, Ş. Limited-Angle Electromechanical Actuator for Micro Wind Turbines Overspeed Protection. In Proceedings of the 2019 11th International Symposium on Advanced Topics in Electrical Engineering (ATEE), Bucharest, Romania, 28–30 March 2019. [CrossRef]
14. Ionica, I.; Modreanu, M.; Boboc, C.; Morega, A. Tridimensional Modeling for a DC, Limited Angle, Torque Motor of Size 16. In Proceedings of the 2016 International Conference and Exposition on Electrical and Power Engineering (EPE 2016), Iasi, Romania, 20–22 October 2016.
15. Analog Devices. Available online: https://www.analog.com/en/products/ltc6993-1.html (accessed on 30 June 2020).

© 2020 by the authors. Licensee MDPI, Basel, Switzerland. This article is an open access article distributed under the terms and conditions of the Creative Commons Attribution (CC BY) license (http://creativecommons.org/licenses/by/4.0/).

Article

Expansion of High Efficiency Region of Wind Energy Centrifugal Pump Based on Factorial Experiment Design and Computational Fluid Dynamics

Wei Li [1,*], Leilei Ji [1], Weidong Shi [2,*], Ling Zhou [1], Hao Chang [1] and Ramesh K. Agarwal [3]

1. Research Center of Fluid Machinery Engineering and Technology, Jiangsu University, No.301, Xuefu Road, Jingkou District, Zhenjiang 212013, China; jileileidemail@163.com (L.J.); lingzhoo@hotmail.com (L.Z.); changhao1514@163.com (H.C.)
2. College of Mechanical Engineering, Nantong University, No.9 Se Yuan Road, Chongchuan District, Nantong 226019, China
3. Department of Mechanical Engineering & Materials Science, Washington University in St. Louis, St. Louis, MO 63130, USA; rka@wustl.edu
* Correspondence: lwjiangda@ujs.edu.cn (W.L.); wdshi@ujs.edu.cn (W.S.); Tel.: +86-137-7555-4729 (W.L.); +86-135-0528-8312 (W.S.)

Received: 24 December 2019; Accepted: 12 January 2020; Published: 19 January 2020

Abstract: The wind energy pump system is a new green energy technology. The wide high efficiency region of pump is of great significance for energy conservation of wind power pumping system. In this study, factorial experiment design (FED) and computational fluid dynamics (CFD) are employed to optimize the hydraulic design of wind energy centrifugal pump (WECP). The blade outlet width b_2, blade outlet angle β_2, and blade wrap angle ψ are chosen as factors of FED. The effect of the factors on the efficiency under the conditions of $0.6Q_{des}$, $0.8Q_{des}$, $1.0Q_{des}$, and $1.4Q_{des}$ is systematically analyzed. The matching feature of various volute tongue angle with the optimized impeller is also investigated numerically and experimentally. After the optimization, the pump head changes smoothly during full range of flow condition and the high efficiency region is effectively improved. The weighted average efficiency of four conditions increases by 2.55%, which broadens the high efficiency region of WECP globally. Besides, the highest efficiency point moves towards the large flow conditions. The research results provide references for expanding the efficient operation region of WECP.

Keywords: wind energy centrifugal pump (WECP); factorial experiment design (FED); optimization; high efficiency area; computational fluid dynamics (CFD)

1. Introduction

The WECP system is a new green energy technology that converts wind energy into electric energy firstly and then uses the battery to drive the pump. At present, the focus of wind power extraction system research is mainly on the performance improvement of wind turbine and the technology of frequency conversion control, whereas the match between pump and system is not well investigated. As wind energy is the variation of solar energy, the wind speed, wind direction, and energy vary with season, altitude, region, and surface coarseness, representing the great change of power generated by wind turbines (the input power of motor). Therefore, the operating condition of wind pump is in variation. The traditional design method of vaned pump will not satisfy the requirements of ultra-wide operating condition of WECP [1–3]. Therefore, it is urgent to expand the high efficiency region of WECPs, improve the efficiency of pump system, and optimize the matching characteristics of volute and impeller, which is of great significance for energy conservation and consumption reduction of wind power pumping system.

Clark R N et al. [4–6] carried out a detailed study on the performance of the small wind power pump and the electric design method of the wind energy pump system. Based on the main composition and principle of the research system, Zhu et al. [7] analyzed the matching optimization of the generator and pump and carried out a test on a wind power pumping system. Lin et al. [8] experimentally simulated the wind power water pumping system and revealed the relationship between flow rate and head power under different wind speeds. However, the focus of above researches is on the performance of wind power pump water system and the matching between generator and pump. It is not specifically designed to improve the performance of pump devices, which could be matched with the wind machinery, especially the performance of the wind power centrifugal pump under the non-design condition. Following are the methods for the performance optimization of centrifugal pumps, namely, loss extreme method, criterion screening method, CFD optimization design method, and experimental design method. The loss extreme method is used to solve the combination of pump geometry parameters that meet the performance to minimize pump losses for optimum performance indicators under the condition of ensuring the flow rate and head of the design working point. Li et al. [9] extended the method of Neumann [10] to the design of all vane pumps. He combined various losses as the minimum principle in the design to ensure the dimensions of components match with each other, analyzed the flow of vane pump under non-design conditions, and proposed the corrective action. Gao and Guo et al. [11] analyzed the influence of impeller and volute design parameters on pump performance, constructed a multi-objective calculation model to solve a set of design parameters by using the highest efficiency and optimal cavitation performance of centrifugal pumps as the objective function, and verified its correctness through instances. Tan et al. [12] analyzed the problems existing in the centrifugal pump hydraulic loss analysis method and gave an expression for the loss coefficient of each component in the form of using specific speed as an independent variable through theoretical analysis. The criterion screening method is proposed for the shortcomings of loss extreme method; the level of optimization depends on whether the criteria are reasonable. Based on the analysis of internal flow mechanism of centrifugal impeller, the criterion screening method established an objective function that reduces losses and achieves performance metrics, sought a combination of geometric parameters of overcurrent components, and screened out the best solution. However, these traditional vane pump optimal operating point design methods cannot meet the requirements of ultrawide operating conditions of WECPs.

CFD optimization is to guide the pump optimization through the simulation of three-dimensional incompressible flow field of pumps by means of high-performance computer [13–16]. In recent years, researchers have developed this method firstly on the preliminary design with one-dimensional theory and then CFD optimization. Goto et al. [17] presented the hydraulic design method of centrifugal impeller based on the whole three-dimensional inverse problem design. Passrucker et al. [18] designed inversely the axial plane projection and blade molded line of impeller by using CFD. Zhou et al. [19] simulated the internal flow of a new type of three-dimensional surface return diffuser to improve the hydrodynamic performance of the deep-well centrifugal pump. Although this method is very convenient to optimize the pump efficiency, it entirely relies on the high-performance computer and experienced technicians using CFD.

Experimental design is a mathematical statistical method for arranging test and analyzing test data. The commonly used experimental design methods are orthogonal test design, FED, response surface design, etc. [20]. The design theory of pump greatly dependent on experience is still not perfect yet. In the optimal design of pumps, the orthogonal test design method is widely adopted. According to the design requirements of non-overload centrifugal pump, Yuan et al. [21] analyzed the effect of geometric parameters on pump efficiency as well as head and shaft power, and they have figured out the main factors based on the orthogonal test. The orthogonal test design of a deep well pump was carried out by Shi et al. [22]. As a result, the influence of design parameters on pump performance was analyzed to find out the significant and nonsignificant factors affecting the pump performance and obtain a better design scheme. Wang et al. [23] carried out a test optimization design for a vortex pump, and the results have been improved greatly. It is proved that the test design is of reference value for the optimization design of pump. The orthogonal test design is easy to obtain the main effect of all factors, but the optimal scheme can only be a combination of the level used in the experiment, and the optimization results will not exceed the range of the taken level. The FED, also known as the full FED, has the advantages of finding main effects from various factors and analyzing the interaction effects between various factors. In medical research, FED is used to select the best treatment scheme and adjust the drug formula, but the optimization design of FED used in the pump is rarely reported [24].

These related studies reveal that these four optimization methods have their own pros and cons. Loss extreme method and criterion screening method are the design methods for the majorization of operating point, which is not applicable for the design of WECPs with wide high efficiency region. Test optimization is accurate and reliable, but it is time-consuming and resource-consuming. CFD optimization is not restricted by physical model and test apparatus, which saves time and money substantially [25–27]. However, due to the lack of standards for judging whether CFD results are accurate or not, it needs to be validated against experimental results. Therefore, in this paper, we propose to combine the FED optimization and CFD optimization together, namely, FED-CFD. By employing numerical simulations and taking the efficiency and head of centrifugal pump under different operating points as the design indexes, the optimal parameter combination is obtained to efficiently match the volute. The optimal design was obtained by FED-CFD method (combination of FED and CFD method) and verified by prototype test. This paper could provide guidance for the optimization of multi condition centrifugal pumps.

2. Hydraulic Design of WECP

2.1. Model Design Requirements

Because of abrupt change of operating conditions of WECP, the primary goal of WECP is to maximize the average efficiency of pump within full range of flow rate conditions; that is, the WECP should have a wider high efficiency region. The traditional design method of pump will not satisfy the requirements of ultrawide operating condition of WECP. When the pump operates under nonoptimized condition, the performance decreases sharply and the constant operation will cause the pump unit instability. Therefore, to improve the overall efficiency of WECP and expand the high efficiency region of WECP, the impeller structure should be optimized based on FED-CFD method. In this paper, the parameters of WECP model are as follows: specific speed $n_s = 68$, designed flow rate $Q = 22$ m^3/h, designed head $H = 11$ m and rotational speed $n = 1450$ r/min.

2.2. Original Parameter Selection

According to the basic pump equation and the infinite blade number hypothesis, the theory head of impeller and the circumferential velocity component of impeller outlet could be expressed as follows,

$$H_{T\infty} = \frac{u_2^2}{g}(1 - \frac{Q}{\pi u_2 D_2 b_2 \psi_2 \eta_V}) \tag{1}$$

$$u_2 = \frac{\pi D_2 n}{60} \tag{2}$$

To keep the original model pump unchanged, the inlet and outlet diameters of impeller are not changed. After transformation, the calculation formula of head is as follows,

$$H_{T\infty} = \frac{1}{g}(\frac{\pi^2 D_2^2}{60^2})n^2 - (\frac{n}{60 g b_2 \psi_2 \eta_V}) \cot \beta_2 nQ \tag{3}$$

The slope of head curve is obtained by Equation (4),

$$\varphi = \frac{n}{60 g b_2 \psi_2} \cot \beta_2 \tag{4}$$

Thus, the pump efficiency could be derived by the above equations as follows,

$$\eta = \frac{\rho g Q H}{P} \tag{5}$$

where φ is the slope of head curve, dimensionless; b_2 is the blade outlet width (mm); β_2 is the blade outlet angle (°); ψ is the blade wrap angle (°); n is the impeller speed (r/min); g is the gravitational acceleration (m/s²); $H_{T\infty}$ is the theoretical head (m); Q is the flow rate (m³/s); D_2 is the outer diameter of the impeller (m); η_v is hydraulic efficiency, dimensionless; and η is pump efficiency, dimensionless. Formula (4) shows that, without changing the inlet and outlet diameter of impeller, the width of blade outlet width b_2, the blade outlet angle β_2, and the blade wrap angle ψ can be used as the adjusting factors to optimize the energy performance of multi-condition of WECP. The definition of impeller key parameters is shown in Figure 1. The axial projection is shown in the left figure and the plane projection is shown in the right figure. The middle streamline a-b-c in the axial projection corresponds to the middle line a'-b'-c' on the blade surface. According to the relevant empirical parameters and combined with the original model hydraulic size, the range of values of these three variables is specified, and the combination of impeller parameters under three different variables is set as the initial factor of FED, as shown in Table 1. Among them, each factor is given two levels, and the code is set according to the level, the low level is set to −1, the high level is set to +1, and the center point is set to 0.

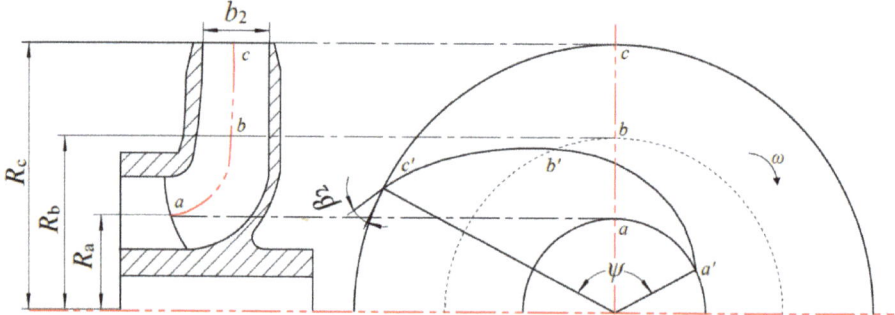

Figure 1. Definition of key parameters.

Table 1. Factor settings.

Independent Variable Factor	−1	0	1
Outlet width of the impeller blade b_2/mm	19	21	23
Outlet angle of the impeller blade β_2/°	16	19	22
Wrap angle of the impeller blade ψ/°	150	160	170

Considering that several factors are implemented simultaneously in the experiment, there may be interactions between factors, so the full FED is selected for analysis [28]. To achieve the principle of "trial repetition", we should arrange repeat tests at the "central point". To ensure repeatability with as few test times as possible, three trials were performed at the center point. A total of 26 (= 23 + 3) trials are conducted. The "center point" of this paper is Impeller 1, Impeller 7, and Impeller 8, respectively. The experiment strategy is shown in Table 2. In the table, A is the blade outlet width b_2, B is the blade outlet angle β_2, and C is the blade wrap angle ψ.

Table 2. Scheduling.

Order of Impeller	Center Point	A	B	C
1	0	0	0	0
2	1	−1	−1	1
3	1	−1	1	−1
4	1	1	−1	−1
5	1	1	−1	1
6	1	−1	1	1
7	0	0	0	0
8	0	0	0	0
9	1	1	1	−1
10	1	1	1	1
11	1	−1	−1	−1

2.3. Hydraulic Design and Establishment of Original Model

The hydraulic design of impeller is carried out by means of PCAD software, and other geometric parameters are kept unchanged during parameter setting. According to the combination of Table 3, the blade outlet width b_2, the blade outlet angle β_2, and the blade wrap angle ψ are designed. The hydraulic model diagram of 11 sets of schemes is designed. Then the 3D geometric model is constructed by using Pro/E. The central point is the hydraulic model of Impeller 1, as shown in Figure 2.

Table 3. Basic geometric parameters of the pump.

Geometric Parameter	Value
Inlet diameter of the impeller D_1 (mm)	62.5
Hub diameter of the impeller D_{hb} (mm)	20
Outlet diameter of the impeller D_2 (mm)	185
Inlet angle of the impeller blade β_1 (°)	14
Outlet angle of the impeller blade β_2 (°)	19
Wrap angle of the impeller blade ψ (°)	160
Number of the impeller blades Z	6
Outlet width of the impeller blade b_2 (mm)	21
Rotate speed n (r/min)	1450
Designed flow Q_{des} (m³/h)	22
Specific speed n_s	68
Designed Head of delivery H (m)	11

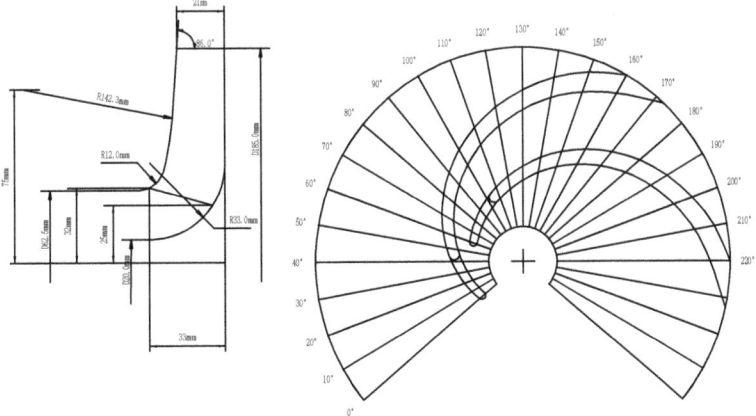

Figure 2. Hydraulic model of Impeller 1.

3. Prediction of Energy Performance and High Efficiency Region

CFD numerical simulation is used to predict the energy performance and high efficiency region of WECP, to select the most optimal scheme. First, the 3D geometric calculation model is constructed, and then the appropriate mesh is generated and imported into ANSYS CFX for the numerical calculation. Although CFD is suitable for simulating the internal flow field of rotating machinery, the numerical settings of CFD should be selected appropriately to ensure the reliability of results. Therefore, a series of numerical calculations for WECP are performed using different grid numbers, turbulence models, and convergence precisions.

3.1. Establishment of the Calculation Domain

Pro/E software has been used to conduct 3D simulation domain of the inlet (semi-spiral suction chamber), impeller, volute, and the front and rear chamber of the model pump, to obtain the 3D mode of whole calculation area of WECP, see Figure 3.

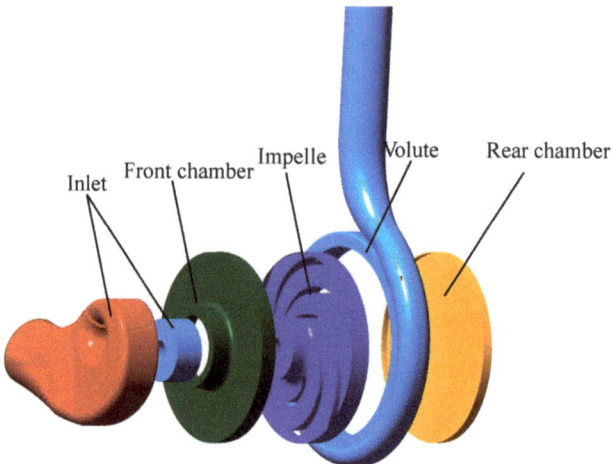

Figure 3. 3D model of the calculation domain.

3.2. Mesh Generation and Mesh Independence

Due to the use of full flow field calculation, the physical size at the gap of orifice is extremely small compared to the physical size of other water bodies; all calculation domains are divided by hexahedral structure meshes. The volute is stretched and combined with O-Block mesh. The spiral suction chamber domain (inlet domain) is divided by the whole block and combined with Y-Block. The impeller is only divided for a single flow channel, and wall boundary layer is achieved by controlling the distribution of nodes at the end of block edge. Then, perform a periodic array to obtain the mesh domain of entire impeller region. The meshing results are shown in Figure 4, and the mesh information of different computing domains is shown in Table 4. As we know, the minimum mesh quality and the minimum angle of grid are the guarantee of simulation accuracy. The minimum mesh quality is over 0.35 and the minimum angle of grid is over 20.7°, which meets the requirement of simulation.

The mesh independence verification has been conducted after mesh generation. The same topology has been applied to each domain, and the mesh number has been changed by adjusting the nodes of each topology line. The mesh quality is still controlled to meet the requirement of simulation. The y+ of each domain is less than 100. Figure 5 shows the pump head with different mesh elements. It shows that the pump head changes little when the mesh number is over 1.1 million elements. After considering the computing resource and the simulation accuracy, the total grid about 1.2 million elements are selected to conduct the simulation.

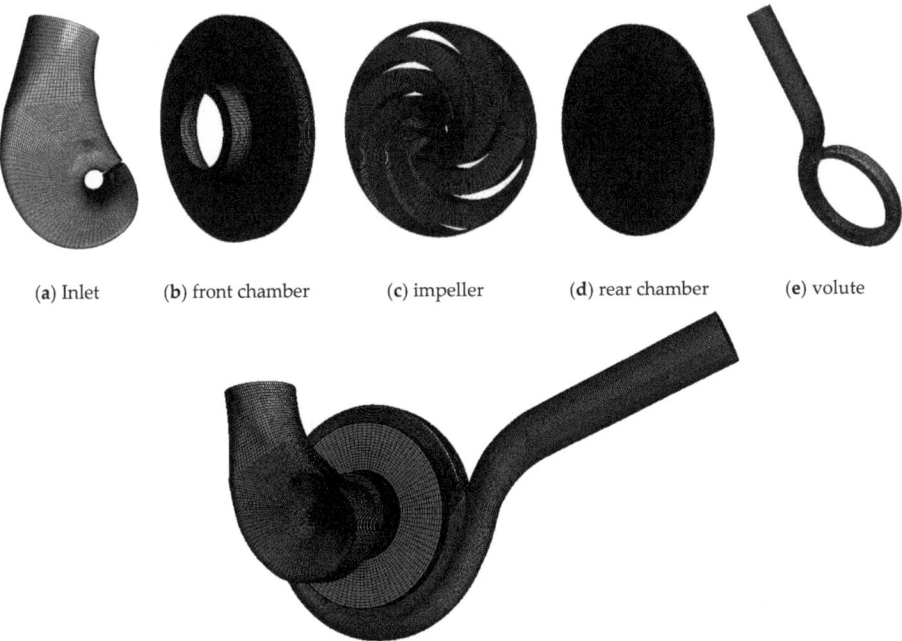

(a) Inlet   (b) front chamber   (c) impeller   (d) rear chamber   (e) volute

(f) Assembled mesh by each part

Figure 4. Structured mesh for the numerical domains.

Table 4. Mesh information of the numerical domain.

Computational Domains	Inlet	Front Chamber	Impeller	Rear Chamber	Volute
Quality	>0.35	>0.72	>0.4	>0.7	>0.55
Minimum Angle/°	>22.6	>54.3	>20.7	>41	>35.2

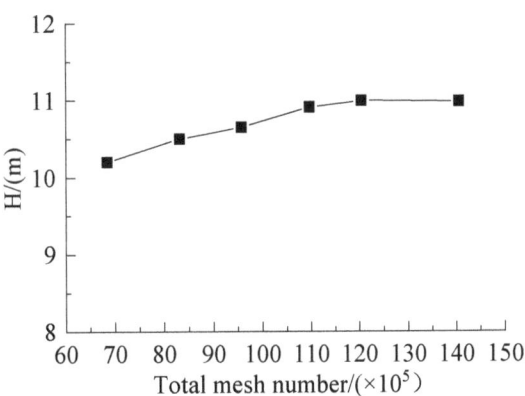

Figure 5. Pump head with different mesh number.

3.3. Selecting the Turbulence Model and Boundary Conditions

A universal turbulence model that is applicable for all flow problems has not yet been formulated; thus, scholars have selected different turbulence models for different turbulent flows. In this study, numerical calculations were performed with ANSYS CFX software, which provides a number of turbulence models. Therefore, five models, i.e., standard k-ε, RNG k-ε, k-ω, and SST k-ω, were selected, and their results were compared with each other. Table 5 shows the numerical results of the pump with different turbulent models under design flow condition of Q_{des} = 22 m^3/h. The relative pump head error from Number 1 to Number 4 is estimated as $(H_1 - H_d)/H_d$ = 5.55%, $(H_1 - H_d)/H_d$ = 3.36%, $(H_1 - H_d)/H_d$ = 6.45% and $(H_1 - H_d)/H_d$ = −0.09%, respectively. The head error is smallest when the standard k-ε model is selected. Therefore, this model was selected in this study. The finite volume method and the second-order windward scheme are applied to discrete the control equations. The inlet is set as velocity inlet and the outlet is set as opening. The standard wall function is chosen to calculate the near wall boundary, and the interfaces close to the impeller are set as Frozen Rotor. The convergence precision is set to 10^{-5}.

Table 5. Performance results with different turbulence models under design flow rate condition.

Number	Turbulence Models	H (m)
1	k-ω	11.61
2	RNG k-ε	11.37
3	SST k-ω	11.71
4	Standard k-ε	10.99

3.4. Identificatioin of High Efficiency Zones

Using the ANSYS CFX post processing tool CFX POST, the performance of the model pumps under the conditions of $0.6Q_{des}$, $0.8Q_{des}$, $1.0Q_{des}$, $1.2Q_{des}$, and $1.4Q_{des}$ are predicted respectively, and the performance parameters of the 11 different schemes are obtained. Figures 6 and 7 show the efficiency curves and head curves for 11 different impellers. To eliminate the influence of volute, the head and efficiency of impeller are calculated. The equations are as follows,

$$H_{impeller} = \frac{p_{2,impeller} - p_{1,impeller}}{\rho g} + \frac{v_{2,impeller}^2 - v_{1,impeller}^2}{2g} \tag{6}$$

$$\eta_{impeller} = \frac{\rho g (Q/3600) \cdot H_{impeller}}{100 P} \times 100\% \tag{7}$$

where $H_{impeller}$ represents the head of impeller, m; $\eta_{impeller}$ represents the efficiency of impeller, %; ρ represents the density of water at room temperature, ρ = 1000 kg/m^3; g represents the gravity, g = 9.8 m/s^2; $p_{1,impeller}$, $p_{2,impeller}$ represent the average circumferential pressure at impeller inlet section and impeller outlet section, respectively, Pa; $v_{1,impeller}$, $v_{2,impeller}$ represents the average circumferential velocity at impeller inlet section and impeller outlet section, m/s; and P represents the power of impeller, kW.

It can be seen from Figures 6 and 7 that either the head or efficiency of the impeller is higher than the pump because there is much more energy expenditure in volute and the influence of volute will discuss in the following section. Moreover, the head or efficiency of each scheme shows a large difference from large flow conditions to small flow conditions. The Impeller 1, Impeller 7, and Impeller 8 are the repeatability tests of center point and the energy performance values of three schemes are basically the same, indicating the stability of test environment. The efficiency and head distribution of other schemes are various at designed flow rate condition. For instance, the highest efficiency occurs at Impeller 2 and the value is about 83.02% while the lowest efficiency occurs at Impeller 4 and the value 79.45%. The highest head occurs at Impeller 9 and the value is 16.86, while the lowest head occurs at

Impeller 2 and the value 13.86. It indicates that the geometry of impeller is significant to the energy characteristics of pump. Thus, it is useful to change the geometry parameters of pump to improve the efficiency.

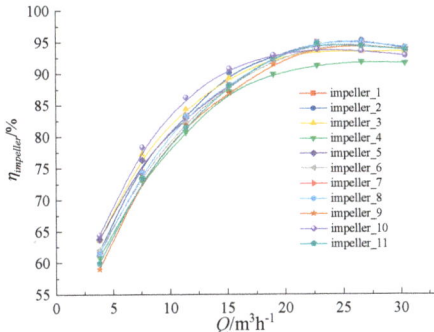

Figure 6. Efficiency curves of impeller under different schemes.

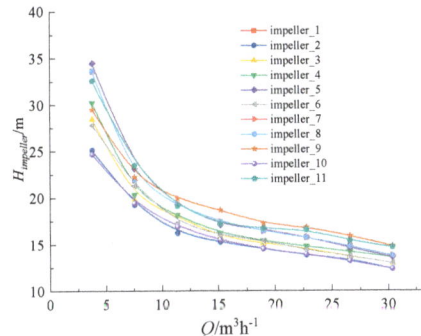

Figure 7. Head curves of impeller under different schemes.

4. Optimal Design of Impeller Based on FED

4.1. Influence of Experimental Factor

The energy characteristics prediction results show that the efficiency of 11 schemes is significantly different under large flow conditions and small flow conditions. Considering the limitation of numerical calculation resources and the actual operating characteristics of WECP, the operating conditions of $0.6Q_{des}$, $0.8Q_{des}$, $1.0Q_{des}$, and $1.4Q_{des}$ are selected as the operating conditions to optimize within the full range of flow rate conditions. The effects of various factors on impeller efficiency under these conditions are analyzed to make the efficiency curve gentler and the high efficiency region wider.

Minitab software is used to analyze the experimental data. All the main and two-order interaction effects are included in the model, including blade outlet width (A), blade outlet angle (B), blade wrap angle (C), and their interaction effects, like the blade outlet width multiply by the blade outlet angle (AB), blade outlet width multiply by the blade angle (AC), blade outlet angle multiply by the blade outlet angle (BC). The interaction above three orders is not considered. Figure 8 is Pareto diagram of the efficiency of impeller parameters.

It can be seen from Figure 8a that the A and B of impeller parameters have a significant influence on the efficiency at $0.6Q_{des}$. According to the intensity of influence, the A is higher than the B. Among the three main effects affecting the efficiency at $0.6Q_{des}$, the A and B are more significant, but the C is not significant. The three 2-factor horizontal interaction effects have no significant effects on the efficiency at 0.6Q_{des}. It can be seen from Figure 8b that the A, B, and C of three main effects have a significant effect on the efficiency of 0.8Q_{des}. Among the three 2-factor horizontal interaction effects, AB has a greater impact on the efficiency of 0.8Q_{des} operating point. The effect of BC and AC on the efficiency under 0.8Q_{des} is not obvious, and it should be removed when the model is improved. It can be seen from Figure 8c that the impeller parameters which have a significant influence on the efficiency of 1.0Q_{des}, including A and AB. As can be seen from the figure, A and AB have a greater impact on efficiency. It can be seen from Figure 8d that all the effects listed in the model have a significant effect on the efficiency of 1.4Q_{des}. According to the intensity of influence, it can be arranged as A, AC, C, B, AB, and BC.

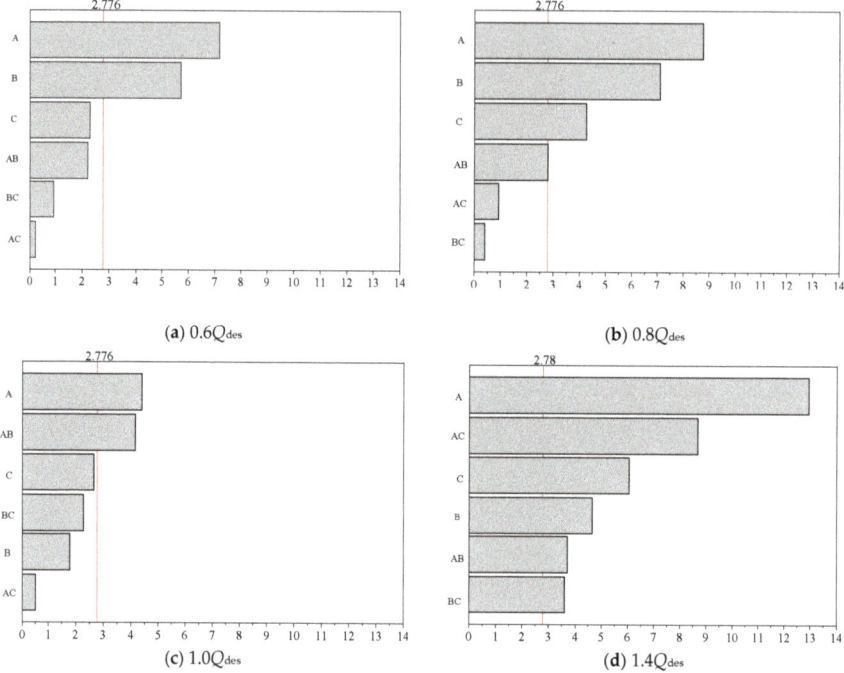

Figure 8. Pareto diagram of efficiency under different operation conditions.

The main effect diagram of efficiency is shown in Figure 9. As can be seen from Figure 9a, the regression lines of blade outlet width and the blade outlet angle are steeper. That is, the two have a greater influence on the efficiency of 0.6Q_{des} operating point. At the same time, the blade wrap angle also has a certain effect on the efficiency. However, the regression line is not as steep as the regression angle of blade outlet width and blade outlet angle: the line is relatively flat. This also confirms that in the Pareto diagram, although the C has not been selected as a significant effect, it is also very close to the critical value. Therefore, it also has a certain impact on efficiency. At the same time, as A and B increase from a low level to a high level, the efficiency decreases under 0.6Q_{des} operating point. It can be roughly seen that in order to maximize the efficiency of 0.6Q_{des} operating point, the A and B can be simultaneously lowered.

It can be seen from Figure 9b that the regression lines of the A, B, and C are steep. That is, the three factors have an obvious effect on the efficiency of 0.8Q_{des} operating point. When A increases from 19 mm to 23 mm or the B increases from 16° to 22°, the efficiency of 0.8Q_{des} operating point shows a downward trend. However, when C increases from a low level to a high level, the efficiency increases at $0.8Q_{des}$. It can be seen from Figure 9c that the A regression line is steep, namely, the A has the most obvious effect on the efficiency of $1.0Q_{des}$ operating point. As the A increases from a low level to a high level, the efficiency of $1.0Q_{des}$ operating point decreases. At the same time, the two factors of B and C also have a certain effect on efficiency. However, the regression line is not as steep as the regression angle of blade outlet width, and it is relatively flat. This also confirms that in the Pareto diagram, the B and C have not been selected for significant effects, but they are also very close to the critical value, so the two have a certain effect on the efficiency. It can be seen from Figure 9d that the regression lines of the A, B, and C are all steep, namely, the three factors have obvious influence on the efficiency of $1.4Q_{des}$ operating point. When the A increases from 19 mm to 23 mm, the efficiency of $1.4Q_{des}$ operating point increases, whereas the B increases from 16° to 22° or the C increases from 150° to 170°, the efficiency of $1.4Q_{des}$ operating point decreases.

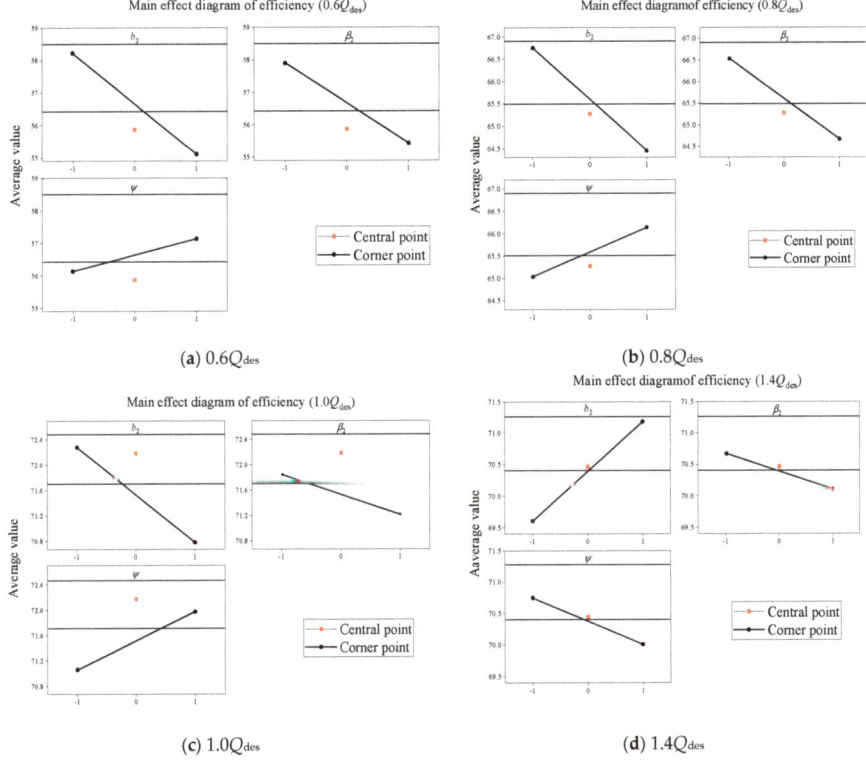

Figure 9. Main effects diagram of different operation conditions.

The interaction effects could reveal the difference in the amount of response between various levels of a factor changes with the different levels of other factors. When there are two or more independent variables in an experimental study, the main effects are not sufficient to find the non-independent effects of several factors. Therefore, the interaction parameters and the interaction effect diagram could evaluate the changes of a factor at different levels which depends on other factors. In this study, the interaction effects show the relationship of different parameters. Figure 10 shows the interaction diagram of efficiency. It can be seen from Figure 10a that the interaction line of BC is almost parallel to the line of AC, which explains that the interaction between these two groups has no significant effect on the efficiency of $0.6Q_{des}$ operating point. The interaction line of AB is not parallel to the other two groups, but the cross trend is not obvious. It can also be seen from the Pareto diagram that, although

AB has not been selected as a significant effect, the *t* value in *t*-test is also very close to the critical line, so it also has a certain impact on efficiency. It can be seen from Figure 10b that the interaction line of AB is obviously not parallel, indicating that the interaction between the two is obvious. Two sets of interaction lines of BC and AC are almost parallel, showing that it has no significant effect on the efficiency of $0.8Q_{des}$ operating point. It can be seen from Figure 10c that the interaction line of AB produces crossover and even form a reverse interaction (when the blade outlet width is at a low level, the efficiency decreases as the blade outlet angle increases, and when the blade outlet width is at a high level, the efficiency increases as the blade outlet angle increases). Similarly, for the BC, the interaction line is not parallel, but not so obvious compared with the interaction line of AB. This also confirms that in the Pareto diagram, BC has not been selected as a significant effect. However, the *t* value in the *t*-test is also very close to the critical value, so it also has a certain impact on efficiency. The interaction line of AC is almost parallel, which has no significant effect on $1.0Q_{des}$ operating point efficiency. It can be seen from Figure 10d that the three sets of interaction lines are not parallel, indicating that the interaction is obvious. Among them, when the blade wrap angle is taken low, the efficiency is not obviously improved as the blade outlet width increases. However, when the blade wrap angle is high, the efficiency increases significantly as the blade outlet width increases. This shows that the AC interaction has a greater impact on efficiency, which is also consistent with the Pareto diagram, and the AC effect is second only to the main effect of A.

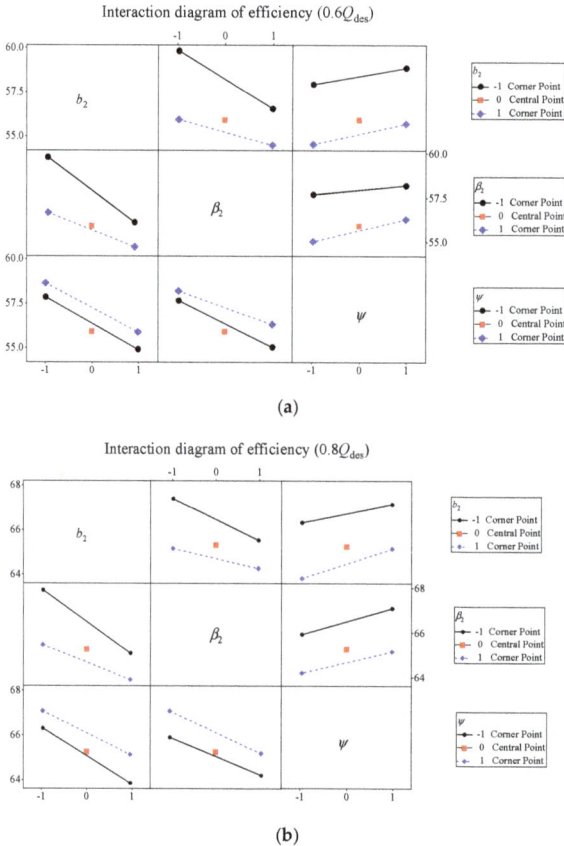

Figure 10. *Cont.*

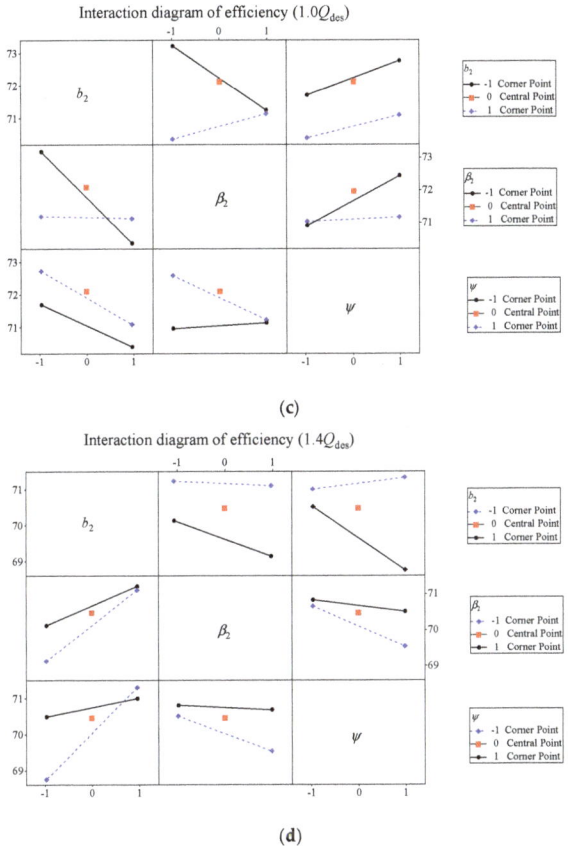

Figure 10. Interaction effect diagram of different flow rate point: (**a**) Interaction effect diagram of $0.6Q_{des}$. (**b**) Interaction effect diagram of $0.8Q_{des}$. (**c**) Interaction effect diagram of $1.0Q_{des}$. (**d**) Interaction effect diagram of $1.4Q_{des}$.

4.2. Geometric Parameter of Optimization Model

Based on variance analysis and residual diagnosis, the regression model of efficiency and impeller parameters under different working operations is obtained:

$$0.6Q_{des}: Y = 56.422 - 1.52A - 1.207B \tag{8}$$

$$0.8Q_{des}: Y = 65.49 - 1.104A - 0.896B + 0.538C + 0.317AB \tag{9}$$

$$1.0Q_{des}: Y = 71.6745 - 0.7412A - 0.3037B + 0.7012AB \tag{10}$$

$$1.4Q_{des}: Y = 70.4436 + 0.775A - 0.28B - 0.3625C + 0.2225AB + 0.52AC - 0.215BC \tag{11}$$

According to the influence of impeller geometric parameters on efficiency, the calculation of the blade outlet width b_2, the blade outlet angle β_2, and the blade wrap angle ψ in Equations (8)–(11) can be performed. In fact, according to the specific requirements of optimization objectives, the target of different working operations is set as "hope big features", namely, Y should be larger at each operating point. Referring to the relevant literature [29], and according to the engineering experience, the weights of the four operating points of $0.6Q_{des}$, $0.8Q_{des}$, $1.0Q_{des}$, $1.4Q_{des}$, are 0.221, 0.286, 0.46, and 0.319, respectively. Afterwards, the response optimization results of blade outlet width b_2, the blade outlet angle β_2, and the blade wrap angle ψ is −0.4902, 0.3307, and 0.5, respectively, which is automatically given by Minitab software. Therefore, a simple interpolation operation could be applied to calculate the geometry parameters. After optimization, the optimal parameters of the impeller according to the target are obtained, where b_2 = 20.02 mm, β_2 = 20°, ψ = 165°. The geometric parameters of the impeller before and after optimization are shown in Table 6.

Table 6. Parameter contrast.

Parameter	Original Impeller	Optimized Impeller
b_2/mm	21	20.02
β_2/°	19	20
ψ/°	150	165

4.3. Energy Performance of the Optimization Model

According to the optimized parameters of the impeller, a new hydraulic diagram of impeller is redrawn with the help of PCAD software. The 3D modeling of impeller was built and the whole flow domain is obtained. The whole fluid domains with optimized impeller are divided into structured meshes and then calculated. The ANSYS CFX is used to calculate the full flow field of the pump before and after optimization. The comparison of the energy characteristic curves of the original impeller and optimized impeller are shown in Figure 11. It can be seen from the figure that the head of the optimized impeller is slightly higher than that of the original impeller from $0.2Q_{des}$ to $1.4Q_{des}$, but the efficiency drops a little. For example, at the designed flow rate condition, the head of original impeller is 10.99, and it is improved to 11.99 after optimization, which is 8.34% higher than the original impeller. However, the expansion of high efficiency region may cost some efficiency of pump, though it is acceptable. If the critical line is set in the efficiency curve, which is 90% of the highest efficiency point, the region of optimized impeller is wider than the original one. Also, from the perspective of weighted average efficiency, the weighted average efficiency of optimized impeller under the condition of $0.6Q_{des}$, $0.8Q_{des}$, $1.0Q_{des}$, $1.2Q_{des}$, and $1.4Q_{des}$ is 69.49%, which is 2.55% higher than the original impeller. It can be seen that the optimized impeller high efficiency region is obviously widened, the impeller efficiency curve is gentle, and the overall efficiency of the WECP is effectively improved, and the optimization objective is achieved.

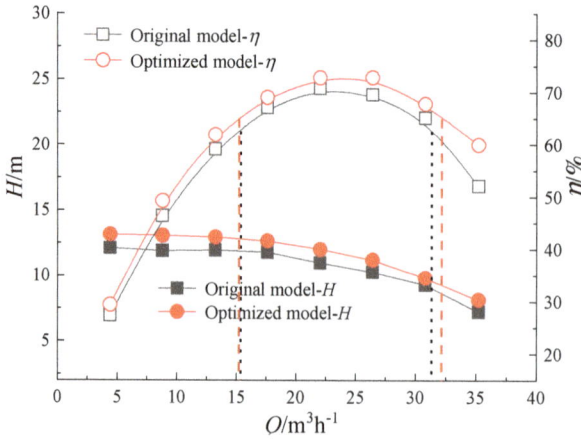

Figure 11. Performance curves of optimized impeller and original impeller.

5. Matching Features between Impeller and Volute

After optimization, the high efficiency region of impeller is obviously widened, which improves the overall efficiency of centrifugal pump and achieves the optimization goal. However, in the experimental design, due to the consideration of test cost, only three impeller parameters, namely, the blade outlet width b_2, the blade outlet angle β_2, and the blade wrap ψ angle, are selected as the independent variables while the geometric parameters of volute are not considered at that stage. However, the impeller and volute are the main overcurrent components of centrifugal pump. Thus, the combination of them also affects the pump performance and there is an optimal matching relationship [30]. To further explore ways to improve the performance index of WECP, and then improve the efficiency of pump unit and achieve energy saving, based on the optimized impeller, by changing the angle of volute tongue angle, the energy characteristics and internal flow of centrifugal pump are studied. The impact provides a theoretical basis for the performance improvement of WECPs.

5.1. Optimum Proposal

Under the circumstance of constant structure parameters of centrifugal impeller, three types of volute are designed with the tongue angle of 23°, 28°, and 33°. The 3D model of different volute domain is shown in Figure 12. The same numerical method is adopted to calculate the whole flow passage considering the clearance in pump cavity. Hexahedral mesh generation method is adopted for all computation areas. After stretching and merging with 3/4 O-Block, the whole O-Block mesh operation is carried out. The same boundary conditions are set in three cases. The inlet boundary is set as velocity inlet and the outlet boundary is set as opening. No-slip boundary condition is applied to all the walls. The standard wall function is used to deal with the flow near the wall region. The dynamic and static interface is set as Frozen Rotor. The computational convergence accuracy is set to 10^{-5}.

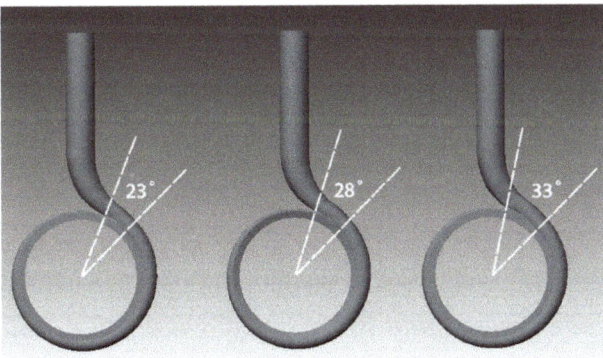

Figure 12. 3D model of different volute.

5.2. Selection of the Volute

5.2.1. Prediction of Energy Characteristics Results

Through numerical simulation, the head curves and efficiency curves of three different types of WECPs are obtained, as shown in Figure 13. Note that the head and efficiency of WECPs are obtained in this section rather than the partial head and efficiency of impeller, due to the influence of volute is considered. It can be seen from the figure that when the angle of the tongue is increased from 23° to 28°, there is little difference in heads under small flow condition. Under high flow conditions, as the flow rate increases, the pump head drops sharply when the tongue is placed at an angle of 23°, and the head curve is flat when the tongue angle is 28°. When the tongue angle changes from 28° to 33°, the head decreases slightly under small flow conditions. After the $1.0Q_{des}$ operating point, as the flow rate increases, the pump head drops more significantly when the tongue angle is 28°. At the $1.6Q_{des}$ operating point, the pump head with the tongue angle of 28° and 33° is 7.94 m and 8.79 m, respectively, with a difference of 0.85 m.

When the tongue angle increases from 23° to 28°, the efficiency does not change much under small flow conditions. Under large flow conditions, when the tongue angle is placed at 28°, the efficiency is significantly higher than that at 23°. When the tongue angle is 28° and 33°, both of the pumps' efficiency does not much different before $1.2Q_{des}$ working condition, but after $1.2Q_{des}$ working condition, with the increase of flow rate, the WECP with the tongue angle of 33° has a more gradual efficiency curve and wider high efficiency zone. The results show that with the increase of tongue angle, the head of WECP has been improved, the high efficiency range of hydraulic efficiency has been widened as well, and the highest efficiency point moves toward the large flow direction.

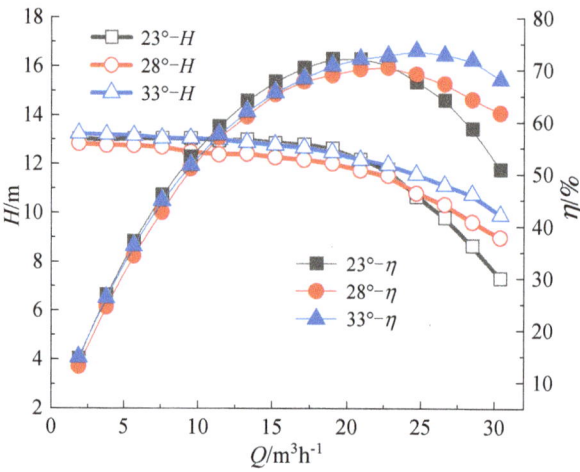

Figure 13. Performance curves with different volutes.

5.2.2. Analysis of Inner Flow

The postprocessing tools CFX-POST provided by ANSYS CFX are utilized to acquire the flow fields of WECP with different tongue angle volutes. The velocity distribution of WECP with different tongue angles on the symmetric surface of volute under design condition is shown in Figure 14. It can be seen from the diagram that the velocity distribution trend on the symmetrical surface of centrifugal pump volutes with different tongue angles are basically the same. The velocity distribution in flow channel near the tongue is not uniform, and there is a clear velocity gradient. Because of the difference of tongue angle, the flow characteristics in the downstream diffuser have been affected greatly. With the tongue angle of 23°, the low speed zone moves backward toward the outlet pipe, and the range and intensity of the low speed zone decrease obviously. However, with the angle of 33°, the influence of tongue angle on the internal flow of centrifugal pump is further weakened. The low velocity area in the diffuser is not obvious and the flow inside the impeller is even more uniform.

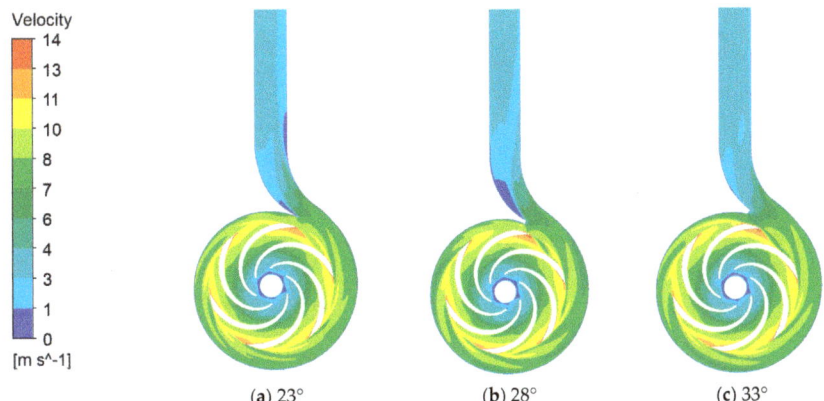

Figure 14. Velocity distribution on volute cross section under design flow rate condition with different volutes.

The pressure distribution of WECP with different tongue angles on the symmetric surface of volute under design condition is shown in Figure 15. The great impact of tongue angle on the pressure distribution at the outlet of volute can be seen in the diagram. With the increase of tongue, the pressure increases at the outlet of volute, which indicates more kinetic energy of fluid are converted to the potential energy. Furthermore, the pressure distribution is more uniform at the outlet of volute when the volute tongue angle is 33°. This shows the better hydraulic feature of volute. Combining with the torque at different tongue angle in Table 7, the torque with different tongue angles have little difference, and it is moderate when the tongue angle is 33°. Those features show the better characteristics of 33° volute.

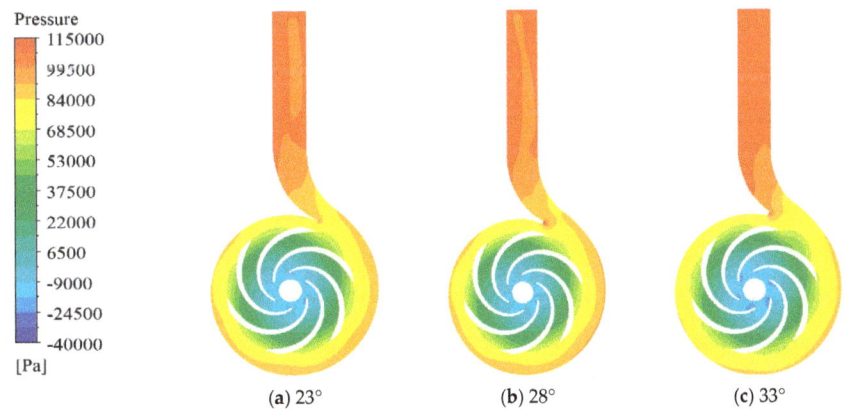

Figure 15. Pressure distribution on volute cross section under design flow rate condition with different volutes.

Table 7. Torque of impeller with different volutes.

Tongue Angle/°	23	28	33
Torque/N·m	6.19	6.11	6.15

Four radial cross sections (Section I (0°), Section III (90°), Section V (180°), and Section VII (270°)) are selected to analyze the internal flow in volute, and the positions of different cross sections are shown in Figure 16. The velocity and the vector in those cross sections are shown in Figure 17, respectively. It can be seen from Figure 17 that the flow inside the volute is unstable, there are various kinds of vortices on each section, and the flow field distribution is complicated. In the section I (0°), the placement of tongue angle has a great influence on the secondary flow velocity distribution. The flow of this section has been affected by the flow of tongue gap, and the mainstream speed is larger. The flow in first section of 33° volute is more stable than the other two types of volute. Close to the tongue area, the vortex has not been formed fully in 0° radial section of 33° volute. At the same time, as the tongue angle reduces, the intensity of vortex is more and more obvious in 0° radial section of 23° volute. A pair of oppositely directed vortices was formed in the radial section and it is clearly visible. In the Section III (90°), more and more fluids enter in the volute. The three types of volutes have opposite vortices in 90° section and exhibit a "right strong left weak" condition. When the fluid reaches section V from III, the fluid entering in the volute further increases. As the impeller outlet flow velocity distribution along the impeller outlet width is not uniform, the fluid entering in the volute is deflected because of this influence. For the 23° volute, the right vortex is not obvious, and the left vortex intensity becomes larger, affecting the velocity distribution of entire section. Although the right vortex still exists in this section of 28° volute, its strength has been weakened and its intensity is lower than the left vortex. The influence of the uneven distribution of impeller outlet velocity of 33° volute is relatively unobvious, and the strength of the left and right vortices is equal. In VII section, the left vortex is still intensive, and the right vortex gradually grows. When the tongue angle is 23°, the intensity of left vortex is slightly higher than the right vortex, which is much different than the other sections. Further, the right vortex in section VII of 28° volute develops much and its strength is nearly equal to the left vortex. When the volute tongue angle grows to 33°, even though the influence range of vortex increases, the left and right vortexes remain basically symmetrical. In summary, the flow fields in volute with different tongue angles present a complex flow state, and the flow advances in the direction of liquid flow almost in the form of a vortex. Along the direction of the liquid flow, as the angle increases, those vortexes also evolve and separate. The placement of tongue angle has a great influence on the vortex flow of each section. As the tongue angle increases, the influence of vortexes on each section becomes weaker, and the flow fields are more stable, which is consistent with the analysis of velocity distribution trend.

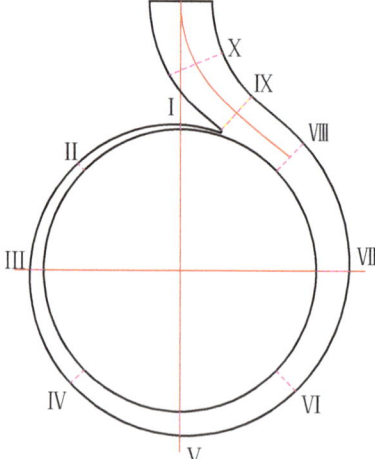

Figure 16. The positions of the cross sections.

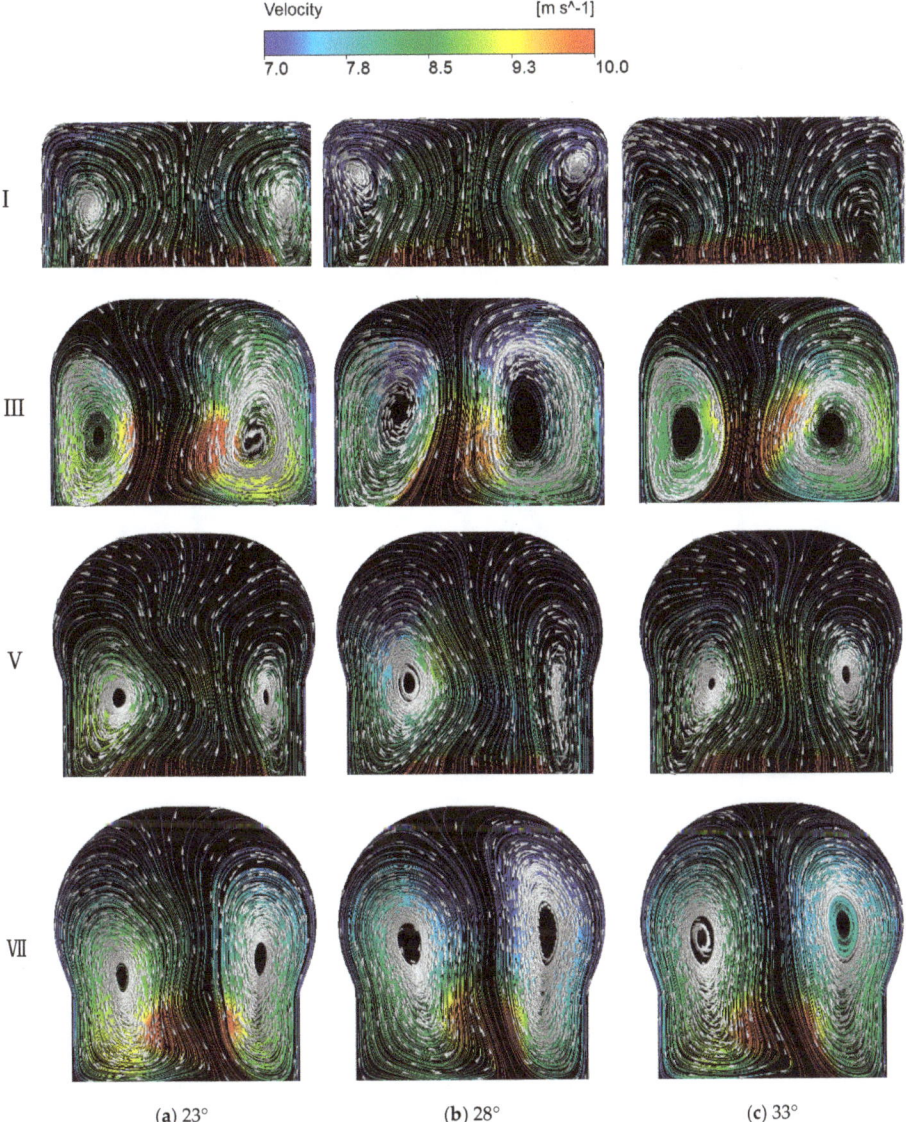

Figure 17. Vector distribution of four radial cross sections under design flow rate condition.

The turbulent kinetic energy distribution inside WECP with different tongue angles under design condition is shown in Figure 18. As can be seen from the diagram, the turbulent kinetic energy distribution inside the spiral water suction chamber, the impeller, and the front and back pump cavity is basically the same, but the turbulent kinetic energy of first section with 23° volute is obviously larger than the other two types of volute. The distribution of turbulent kinetic energy inside the volute cross section with the volute tongue angle of 33° is the most uniform with the tongue angle of 33°. Through the analysis of the internal flow fields of WECP, it is known that the proper increase of tongue angle can effectively improve the internal flow of WECPs.

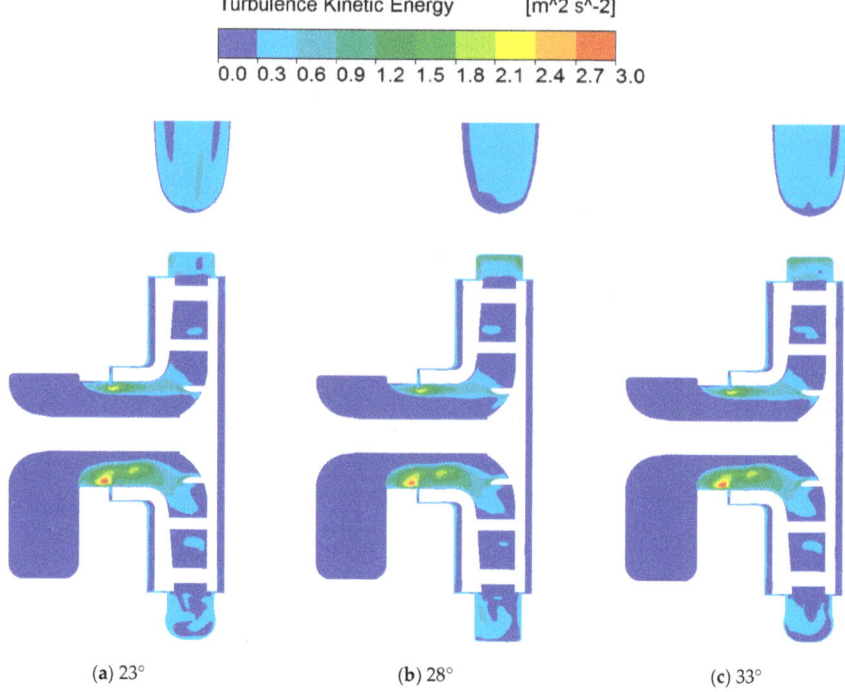

Figure 18. Distribution of turbulent kinetic energy under design condition with different tongue angles.

6. Verification of Energy Performance of Optimized Model

6.1. Testing Pump and Facilities

The impeller and volute are made of fully transparent plexiglass. The physical model of impeller and volute is shown in Figure 19. To verify the accuracy of numerical simulation, the test bed has been established for testing energy characteristic of WECP as shown in Figure 20. Its test bed reaches 1 level required precision. The external characteristic test has been carried out on the prototype model of WECP with the tongue angle of 33°.

Figure 19. Model.

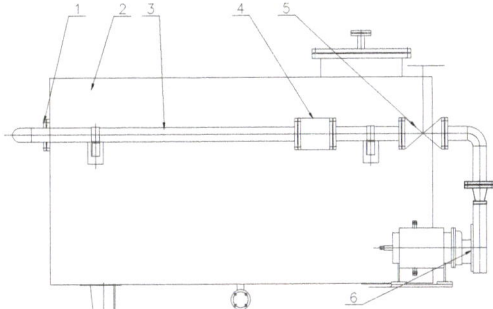

Figure 20. Test bed. 1. Pipeline export. 2. Water tank. 3. Pipeline. 4. Electromagnetic flowmeter. 5. Discharge valve. 6. Test pump.

6.2. Analysis of Energy Characteristic Results

During the test of energy characteristics, the reliability can be guaranteed by collecting the data thrice under each flow rate condition and the comparison between repeated results are is shown in Figure 21. According to the repeatability test results, the low difference in head and efficiency of three repeated tests indicates that the test environment is stable and the reliability of test bed and test system is good.

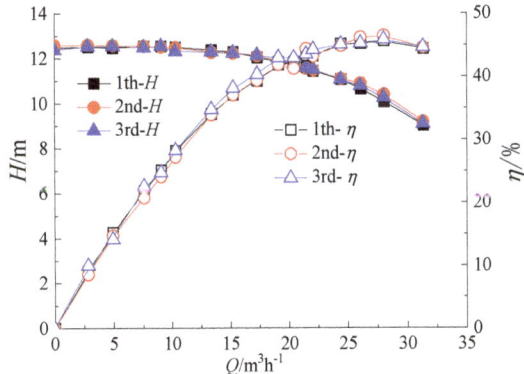

Figure 21. Performance curves of repetitive testing.

Comparing experimental and CFD results, the external characteristic curves are plotted respectively, as shown in Figure 22. Even though the pump efficiency of WECP is calculated by the same equation, the actual value of efficiency is impossible to be the same because of the complexity of boundary in reality. In engineering fields, the difference is acceptable as long as the error is within a reasonable range. Therefore, the difference of head is particularly discussed in this paper. It can be seen from the figure that the pump head of numerical calculation is close to the experiment results and the calculated head is slightly higher than the test. At the design point, the calculation head is 11.89 m and the test head is 11.28 m, which indicates the pump head is predicted to be about 5.4% higher than the test value. However, under the small flow condition, the test value is close to the calculated value. Furthermore, there is a certain difference under large flow condition. The average head of numerical calculation is higher than the experimental value of ~5%. Excluding some small influence factors, two main reasons may account for this phenomenon. On one hand, the inlet flow is simplified into uniform flow in numerical calculation, which is different from the actual situation. On the other hand, the head will be reduced by cavitation occurring under large flow rate condition, which is not considered in numerical calculation. On the whole, the numerical simulation head is basically same as the experimental measurement head. The results show that the numerical calculation method adopted in this paper is accurate and feasible as a means of experimental implementation.

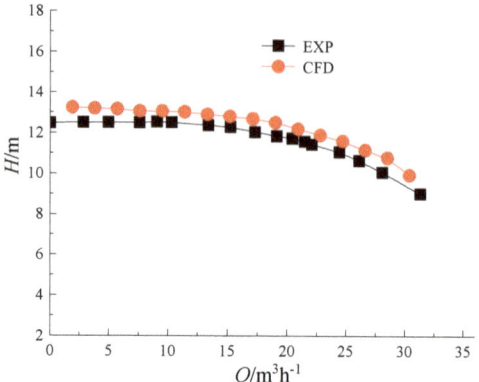

Figure 22. Comparison between experiment and simulation.

7. Conclusions

In this paper, based on FED and CFD methods, the optimal design of WECP has been successfully conducted. Eleven schemes of experiments were designed to research the effect of experimental factors, including blade outlet width b_2, blade outlet angle $β_2$, and blade wrap angle $ψ$. A mathematical model related to the efficiency of different working conditions and the main geometric parameters of impeller was established. Meanwhile, the matching characteristics of impeller and volute were numerically studied. The results of the analysis are as follows.

(1) The relationship model between impeller geometric parameters and efficiency can be established by FED-CFD method. The mathematical model contains the main effects and interaction effects of impeller geometric parameters on pump efficiency. Blade outlet width dominates the main effect and interaction term. The interaction between the blade outlet width and the blade wrap angle or the interaction between blade outlet width and the blade outlet angle has a significant effect on the efficiency of large flow operating point. However, the effect of interaction on efficiency in small flow conditions can be ignored.

(2) The matching of impeller and the volute is essential to improve the overall efficiency. Under large flow conditions, the volute tongue angle is positively correlated with head and efficiency. As the tongue angle is increased, the head of centrifugal pump is increased, the efficient range of hydraulic efficiency is widened, and the highest efficiency point is moved toward the large flow direction.

(3) Implement the optimal design of WECP by FED-CFD method. After optimization, the head curve of impeller is relatively flat, the efficiency is at a high level, and the high efficiency region is widened. The weighted average efficiency of four operating points increases by 2.55%, which improves the overall efficiency of WECP. Through the experimental verification, the established optimization design method for combination of FED and CFD is reliable and effective, and can be applied to the optimization design of multi-condition centrifugal pump.

Author Contributions: Methodology and project administration, W.L.; writing—original draft preparation and review and editing, L.J.; review and editing, W.S., L.Z., H.C. and R.K.A. All authors have read and agreed to the published version of the manuscript.

Funding: The work was sponsored by the National Key R&D Program Project (No. 2017YFC0403703), National Natural Science Foundation of China (Nos. 51679111, 51409127 and 51579118), PAPD, Key R&D Program Project in Jiangsu Province (BE2016319, BE2017126), Natural Science Foundation of Jiangsu Province (Nos. BK20161472, BK20160521), Science and Technology Support Program of Changzhou (No. CE20162004), Key R&D Program Project of Zhenjiang (No. SH2017049), and Scientific Research Start Foundation Project of Jiangsu University (No. 13JDG105), Postgraduate Research & Practice Innovation Program of Jiangsu Province (KYCX19_1601).

Conflicts of Interest: The authors declare no conflict of interest.

References

1. Wenye, W.; Yongguang, H.; Shuo, Y.; Kangqian, M.; Xiaoyong, Z.; Pingping, L. Optimal design of wind machine impeller for frost protection based on CFD and its field test on airflow disturbance. *Int. J. Agric. Biol. Eng.* **2015**, *8*, 43–49.
2. Zhao, R.; Li, H.; Zhang, D.; Huang, J.; Shi, W. Numerical investigation of pump performance and internal characteristics in ALIP with different winding schemes. *Int. J. Appl. Electromagn. Mech.* **2018**, *57*, 39–51. [CrossRef]
3. Fangwei, X.; Rui, X.; Gang, S.; Cuntang, W. Flow characteristics of accelerating pump in hydraulic-type wind power generation system under different wind speeds. *Int. J. Adv. Manuf. Technol.* **2017**, *92*, 189–196. [CrossRef]
4. Nelson, V.; Clark, R.; Foster, R. *Wind Water Pumping*; West Texas A&M University: Canyon, TX, USA, 2004; p. 108.
5. Vick, B.D.; Clark, R.N. Performance of wind-electric and solar-PV water pumping systems for watering livestock. *J. Sol. Energy Eng.* **1996**, *118*, 212–216. [CrossRef]
6. Muljadi, E.; Bergey, M.; Flowers, L.; Green, J. Electric design of wind-electric water pumping systems. *J. Sol. Energy Eng.* **1996**, *118*, 246–252. [CrossRef]
7. Yanxiang, Z.; Zhizhang, L.; Tianqiang, H.; Maofeng, G. Research on wind pumping system and parameter matching. *Renew. Energy* **2006**, *23*, 67–69.
8. Lin, Z.-X.; Huo, T.-Q. Experimental study on water system of wind power pump. *J. Inn. Mong. Univ. Technol. Nat. Sci. Ed.* **2009**, *28*, 139–141.
9. Li, S.-H. *Non-Design Conditions of Vane Pump and Its Optimal Design*; Mechanical Industry Press: Beijing, China, 2006.
10. Neumann, B. *The Interaction between Geometry and Performance of a Centrifugal Pump*; Mechanical Engineering Publications: Beijing, China, 1991.
11. Gao, J.-Y.; Guo, Z.-M. Optimization research on geometric parameters of centrifugal pump impeller and spiral case design. *Pump Technol.* **2007**, *4*, 6–9.
12. Tan, M.-G.; Liu, H.-L.; Yuan, S.-Q. Calculation of hydraulic loss of centrifugal pump. *J. Jiangsu Univ. (Nat. Sci. Ed.)* **2007**, *5*, 11.
13. Korakianitis, T.; Hamakhan, I.A.; Rezaienia, M.A.; Wheeler, A.P.; Avital, E.J.; Williams, J.J. Design of high-efficiency turbomachinery blades for energy conversion devices with the three dimensional prescribed surface curvature distribution blade design (CIRCLE) method. *Appl. Energy* **2012**, *89*, 215–227. [CrossRef]
14. Ouchbel, T.; Zouggar, S.; Elhafyani, M.L.; Seddik, M.; Oukili, M.; Aziz, A.; Kadda, F.Z. Power maximization of an asynchronous wind turbine with a variable speed feeding a centrifugal pump. *Energy Convers. Manag.* **2014**, *78*, 976–984. [CrossRef]
15. Ji, L.; Li, W.; Shi, W. Influence of different blade numbers on unsteady pressure pulsations of internal flow field in mixed-flow pump. *J. Drain. Irrig. Mach. Eng.* **2017**, *35*, 666–673.
16. Yang, M.; Wang, D.; Gao, B.; Lu, S. Influences of guide vane-casing volute positions on performance of nuclear reactor coolant pump. *J. Drain. Irrig. Mach. Eng.* **2016**, *34*, 110–114.
17. Goto, A.; Nohmi, M.; Sakurai, T.; Sogawa, Y. Hydrodynamic design system for pumps based on 3-D CAD, CFD and inverse design method. *ASME J. Fluids Eng.* **2002**, *124*, 329–335. [CrossRef]
18. Passrucker, H.; Van den Braembussche, R.A. Inverse design of centrifugal impellers by simultaneous modification of blade shape and meridional contour. In *ASME Turbo Expo 2000: Power for Land, Sea, and Air*; American Society of Mechanical Engineers: New York, NY, USA, 2000.
19. Zhou, L.; Shi, W.; Lu, W.; Hu, B.; Wu, S. Numerical investigations and performance experiments of a deep-well centrifugal pump with different diffusers. *ASME J. Fluids Eng.* **2012**, *134*, 071102. [CrossRef]

20. Mu, X.-F. *Research and Application of Multidisciplinary Design Optimization Agent Model Technology*; Nanjing Aerospace University: Nanjing, China, 2004.
21. Yuan, S.-Q.; Cao, W.-L. Design of non-overload centrifugal pump by orthogonal test. *Pump Technol.* **1991**, *3*, 1–16.
22. Shi, W.; Zhou, L.; Lu, W.; Zhang, L.; Wang, C. Orthogonal test and optimization design of high lift deep well centrifugal pump. *J. Jiangsu Univ. (Nat. Sci. Ed.)* **2011**, *32*, 400–404.
23. Chunlin, W.; Haibo, P.; Jian, D.; Binjuan, Z.; Fei, J. Optimization design of swirling pump based on response surface method. *J. Agric. Mach.* **2013**, *44*, 59–65.
24. Cheng, J.-L.; Zhen, M.; Lou, J.-Q. Comparison of common experimental optimization design methods. *Lab. Res. Explor.* **2012**, *7*, 7–11.
25. He, X.H.; Cai, S.C.; Deng, Z.D.; Yang, S. Experimental and numerical study of flow characteristics of flat-walled diffuser/nozzles for valveless piezoelectric micropumps. *Proc. Inst. Mech. Eng. Part C J. Mech. Eng. Sci.* **2017**, *231*, 2313–2326. [CrossRef]
26. Kang, C.; Mao, N.; Pan, C.; Zhu, Y.; Li, B. Effects of short blades on performance and inner flow characteristics of a low-specific-speed centrifugal pump. *Proc. Inst. Mech. Eng. Part A J. Power Energy* **2017**, *231*, 290–302. [CrossRef]
27. Kai, W.; Zixu, Z.; Linglin, J.; Houlin, L.; Yu, L. Research on unsteady performance of a two-stage self-priming centrifugal pump. *J. Vibroeng.* **2017**, *19*, 1732–1744. [CrossRef]
28. He, Z.; Pan, Y. Factor test, comparative study of RSM and Taguchi method. *Mech. Des.* **1999**, *16*, 1–4.
29. Wang, K. *Hydraulic Design and Optimization of Centrifugal Pump under Multiple Working Conditions and Its Application*; Jiangsu University: Jiangsu, China, 2011.
30. Weidong, S.; Pingping, Z.; Desheng, Z.; Ling, Z. Unsteady flow pressure fluctuation of high-specific-speed mixed-flow pump. *J. Agric. Eng.* **2011**, *27*, 147–152.

© 2020 by the authors. Licensee MDPI, Basel, Switzerland. This article is an open access article distributed under the terms and conditions of the Creative Commons Attribution (CC BY) license (http://creativecommons.org/licenses/by/4.0/).

MDPI
St. Alban-Anlage 66
4052 Basel
Switzerland
Tel. +41 61 683 77 34
Fax +41 61 302 89 18
www.mdpi.com

Energies Editorial Office
E-mail: energies@mdpi.com
www.mdpi.com/journal/energies

www.ingramcontent.com/pod-product-compliance
Lightning Source LLC
LaVergne TN
LVHW070044120526
838202LV00101B/427